# Validation of Biopharmaceutical Manufacturing Processes

ACS SYMPOSIUM SERIES **698**

# Validation of Biopharmaceutical Manufacturing Processes


**Brian D. Kelley,** EDITOR
*Genetics Institute*

**R. Andrew Ramelmeier,** EDITOR
*Merck & Co., Inc.*


Developed from a symposium sponsored by the Division of Biochemical Technology, at the 213th National Meeting of the American Chemical Society, San Francisco, California, April 13–17, 1997


American Chemical Society, Washington, DC

**Library of Congress Cataloging-in-Publication Data**

Validation of biopharmaceutical manufacturing processes / Brian D. Kelley, R. Andrew Ramelmeier.

p. cm.—(ACS symposium series, ISSN 0097–6156; 698)

"Developed from a symposium sponsored by the Division of Biochemical Technology at the 213th National Meeting of the American Chemical Society, San Francisco, California, April 13–17, 1997."

Includes bibliographical references and indexes.

ISBN 0–8412–3567–8

1. Pharmaceutical biotechnology—Congresses.

I. Kelley, Brian D., 1964– . II. Ramelmeier, R. Andrew. III. American Chemical Society. Division of Biochemical technology. IV. American Chemical Society. Meeting (213th : 1997: San Francisco, Calif.) V. Series.

RS380.V34 1998
615′.19—dc21

98–18617
CIP



The paper used in this publication meets the requirements of American National Standard for Information Sciences—Permanence of Paper for Printed Library Materials, ANSI Z39.48–1984.



Distributed by Oxford University Press



PRINTED IN THE UNITED STATES OF AMERICA

# Advisory Board

**ACS Symposium Series**

# Foreword

THE ACS SYMPOSIUM SERIES was first published in 1974 to provide a mechanism for publishing symposia quickly in book form. The purpose of the series is to publish timely, comprehensive books developed from ACS sponsored symposia based on current scientific research. Occasionally, books are developed from symposia sponsored by other organizations when the topic is of keen interest to the chemistry audience.

Before agreeing to publish a book, the proposed table of contents is reviewed for appropriate and comprehensive coverage and for interest to the audience. Some papers may be excluded in order to better focus the book; others may be added to provide comprehensiveness. When appropriate, overview or introductory chapters are added. Drafts of chapters are peer-reviewed prior to final acceptance or rejection, and manuscripts are prepared in camera-ready format.

As a rule, only original research papers and original review papers are included in the volumes. Verbatim reproductions of previously published papers are not accepted.

ACS BOOKS DEPARTMENT

# Contents

# Preface

The papers published in this volume were drawn from a session entitled *Validation of Biopharmaceutical Manufacturing Processes* held at the 1997 American Chemical Society National Meeting in San Francisco. The session was sponsored by the Division of Biochemical Technology. Attendance at the session was strong throughout the day, as eighteen papers were presented on many aspects of process validation. We believe this reflects the importance of validation in the licensure and manufacturing of biopharmaceuticals, including recombinant and non-recombinant proteins, vaccines, agents for genetic therapies, and carbohydrate-based drugs.

These therapeutic agents and vaccines originate from biological processes, which may lead to a complex mixture of product isoforms; often, these are poorly characterized in comparison to low molecular weight compounds. This combination of biological complexity and variability during product synthesis and purification coupled with the difficulty in fully characterizing the product composition motivates extensive process validation of these processes. While the Food and Drug Administration and its international counterparts publish guidance on validation, these guidelines are general and require considerable interpretation.

Various responses are possible given this incompletely defined mandate. Companies may seek to minimize the investment in validation studies, by identifying the minimally acceptable validation package. Alternatively, a validation master plan can grow to enormous proportions, as all possible studies are conducted in an effort to answer every conceivable validation question. Companies must remain focused on the primary motivation for completing a process validation package, which is to demonstrate both an understanding of the manufacturing process and how to control the process so that product of sufficient quality and yield is produced in a consistent manner. The application of sound scientific, engineering, and statistical methods are necessary to complete this task.

The intent of this volume is to provide a forum for the presentation of various components of a process validation package. The papers address all aspects of biopharmaceutical manufacturing processes, including cell culture and fermentation, product purification, and fill-finish operations. A symposium series book of this size cannot possibly convey information on all topics required for complete process validation; however, we hope that this book is useful as a reference which provides insights into the design and execution of process validation studies from a number of companies, many of which hold licenses for the manufacture of biotherapeutics and vaccines.

**Acknowledgments**

The editors thank all those who made oral presentations at the ACS meeting, and especially those who contributed manuscripts for this book. We recognize the sensitive nature of preparing material on process validation to be shared with a larger audience, and are deeply grateful to those who persevered through it all. We also thank the reviewers, whose efforts have assured that the chapters in this book are of the highest technical and compositional quality.

BRIAN D. KELLEY
Genetics Institute, Inc.
Andover, MA 01810

R. ANDREW RAMELMEIER
Merck Research Laboratories
West Point, PA 19486

## Chapter 1

# Historical, Current, and Future Trends for Validating Biological Processes

**R. Andrew Ramelmeier[1], Brian D. Kelley[2], and Christi Van Horn[3]**

**[1]Merck Research Laboratories, West Point, PA, 19486**
**[2]Genetics Institute, Inc., Andover, MA, 01810**
**[3]Merck Manufacturing Division, West Point, PA 19486**

Since the early 1970's, the FDA has stressed the importance of process validation for manufacturing drugs and medical devices. The emphasis was placed first on sterilization validation and aseptic processing. More recently, however, the emphasis has shifted towards all aspects of the manufacturing process. The FDA has established specific requirements in the CFR and issued guidelines (1), but the interpretation of the validation requirements, especially for biological products, can vary. This is attributed, in part, to rapidly advancing technologies in the biotechnology industry. To facilitate the "validation process" in the biotech industry, some standardization is clearly required.

Emphasis on rigorous process validation in biotechnology has lagged behind the rest of the pharmaceutical industry. This is in part due to the recent introduction of many bio-pharmaceuticals, but also to their significant complexity compared to their non-biological counterparts. Most of what has been learned for the pharmaceutical industry can be and has been applied for bioprocesses. This introduction reviews these principles and also highlights the special challenges associated with processes in the biopharmaceutical industry. The chapters in this book focus primarily on the validation of bulk biopharmaceutical processes (and not on the other elements of a validation program, such as raw materials, equipment and facility qualification or assay and cleaning validation). The latter are described in great detail in other texts (2-5). The approaches discussed in this introduction originate from the authors' experiences, as well as those presented or published by others in the industry (2,3,6,7).

## Background

**Incentive for validating a biopharmaceutical manufacturing process.** Those who do not fully grasp the fundamental significance of process validation often view it as tedious, repetitive and time consuming. So, why do it? The most compelling reason is that validation makes good scientific and engineering sense. Other industries (aerospace, automotive, and cosmetic, for example) have long embraced the concept of validation to assure safe and high quality products. If experiments, analytical assays, equipment operations, vendor evaluations, process scale-up and other related activities are completed and documented properly from process development through production start-up, then the validation package is developed with the process. In this way, the quality assurance and FDA compliance goals are achieved as well. A solid validation strategy can also result in significant economic benefit as lot rejection or recall can be prevented and less troubleshooting will be required. In addition, early consideration of validation requirements and development of a validation plan can save a company time and money by preventing costly delays.

**Definitions** As many organizations and companies have adopted their own "language" for describing the validation process, the authors felt the need to define the key terms and concepts that will be used throughout this chapter and this book. Consistent and proper use of these terms and concepts will reduce confusion when discussing process validation. The key terms are as follows:

**Process validation** - establishing documented evidence which provides a high degree of assurance that a specific process will consistently produce a product meeting its pre-determined specifications and quality characteristics (1).

**Prospective Validation**- establishing documented evidence that a system does what it purports to do based on a pre-planned protocol (9). Often, these activities are performed with scaled-down systems which accurately mimic the full-sized production process. This approach is discussed in more detail in the section on Current Trends in this chapter.

**Concurrent Validation** - establishing documented evidence that a system does what it purports to do based on information generated during actual implementation of the process (9).

**Retrospective Validation** - establishing documented evidence that a system does what it purports to do based on a review and analysis of historical information (9).

**Critical Process Parameters** - the important process variables which affect the process outcome.

**Critical Quality Attributes** - the corresponding critical measures of process outcome or performance.

**Scaled-down model** - an accurate representation of the full-scale production process at a smaller scale. Generally, before a scaled-down model is used for validation studies, the model must be qualified with respect to performance at full-scale.

**Worst-case** - the conditions where upper and lower processing limits and circumstances, including those within standard operating procedures, which pose the greatest chance of process or product failure when compared to ideal conditions (1). This condition, however, should not result in process failure.

**FDA Guidelines**: The most recent guidelines on process validation were issued in 1987 (1). In this document the FDA described several key concepts including quality assurance, the definitions of process validation and worst-case testing, the proper documentation of development data, and the use of prospective versus retrospective approaches. The concepts are broadly stated and have general applicability, as the great variety of products, processes and equipment prevent summarizing the specific validation elements in one document.

To provide adequate quality assurance, a validation program must be coupled with appropriate in-process controls and routine end-product testing. The FDA states that "validation and end-product testing are not mutually exclusive". Validation will reduce the amount of in-process and end-product testing, but may not eliminate them completely.

Manufacturers must prepare a written validation plan or protocol which specifies the procedures and tests to be conducted and the data to be collected and analyzed. The testing should include in-process monitoring of key process variables. In the event that testing cannot adequately measure certain attributes, "process validation should be derived primarily from qualification of each system used in production and from consideration of the interaction of the various systems". The FDA further states that this protocol should specify a sufficient number of replicate process runs to demonstrate reproducibility and to provide an accurate measure of variability among successive runs. The FDA emphasizes that these test conditions for these runs should encompass "worst case" or "most appropriate challenge" conditions as defined above. The FDA obviously does not expect all the variables to be challenged to their extremes; minimally, those critical variables that affect product quality should be subjected to a stressed condition (short of failure) to demonstrate robustness and define process boundaries. This requirement is difficult to fulfill when validating bulk manufacturing processes, which are comprised of several steps with more than one critical parameter; three lots may not be able to bridge all variables. The required approach will most like involve the use of multi-factorial experimental design.

For this reason, the FDA realizes that under some circumstances a rigorous prospective validation approach is not appropriate (for example, non-aseptic, bulk processes well removed from the finishing steps) and that some post-market validation can be applied. Chapman summarizes the use of the Proven Acceptable Range (PAR) approach using validation data from routine batch records (8). This data should be augmented with prospective laboratory or pilot plant work to insure that the process is sufficiently qualified. Under no circumstances, however, should a product be released for clinical use or commercial sale without sufficient assurance of the product's "fitness-for-use".

## Historical Perspective

The FDA first defined the term "validation" in the 1970's in response to sterility problems in the large volume parenteral industry. The FDA used "validation" to describe new requirements they felt the pharmaceutical industry must follow to obtain higher sterility assurance in the preparation of sterile components, equipment, and products. The initial focus was sterilization validation, including steam autoclaves, dry heat ovens, filtration, and ethylene oxide and radiation treatments. These operations were based on physical laws and microbial inactivation kinetics. Since this area of validation involves discreet systems and tasks, it was rapidly adopted throughout the industry. One could not dispute the value of solid mathematical, engineering, and scientific principles when conducting these studies; they have led to a much lower incident of sterility breeches. Consequently, sterilization validation has been embraced by the pharmaceutical industry for 20 years now as standard practice.

The definition of validation expanded in the 1980s, but was still heavily weighted towards aseptic processing. Sterilization validation activities were adopted first in active-ingredient manufacturing and then applied to non-sterile finished pharmaceutical dosage forms. In 1980, aseptic process validation was expanded to include media fills, environmental monitoring, and disinfection/sanitization efficacy testing. These new disciplines were primarily associated with steps following sterile filtration, where many aseptic manipulations of sterilized individual components can occur. Validation issues were expanded again in the 1980s to include pharmaceutical-grade water systems.

In 1984, the FDA first published the Guideline on General Principles of Process Validation. It was clear at that time that the validation principles had finally spilt-over to the non-sterile dosage forms and would be strictly enforced. Analytical methods validation had already been defined in the United States Pharmacopoeia and computerized systems were being targeted during cGMP inspections, as well as being the subject of guidance documents from the FDA. During this decade, the FDA prepared several such guidelines, intended as training tools for new FDA inspectors. These guidelines provided some preliminary requirements for validation and process controls, which companies began utilizing to build their validation programs. Finally, in 1987 the FDA issued a more comprehensive (and the most recent) version of the Guideline on General Principles of Process Validation as well as the Guideline on Sterile Drug Products Produced by Aseptic Processing.

By 1990, validation was also being applied by the manufacturers of active-pharmaceutical ingredients and biotechnology-derived products. In 1991, the FDA published two milestone guidelines: Guide to Inspection of Bulk Pharmaceutical Chemicals and Guideline for the Inspection of Biotechnology Manufacturing Facilities. The latter document acknowledged that biotechnology-derived products are complex molecular entities made by biological processes, but also provided criteria against which biotech facilities would be inspected by the FDA. Whether inspected by the Center for Drug Evaluation and Research (CDER) or the Center for Biologics Evaluation and Research (CBER), the cGMP and validation requirements are the same. This inspector training guide addresses validation as is expected today and its impact on manufacturing equipment/systems, test methods, raw materials, and product distribution. When embraced by a company, validation has proven to be a key element in expediting regulatory approval and increasing productivity and opportunities for cost savings. As such, it has become an essential business strategy with the potential to provide clear competitive opportunities to achieving market share, earlier product launches, speedy regulatory filing approval, and a reliable supply to market of high quality products.

## Validation - Retrospective, Concurrent, or Prospective

Process validation is not a department or a document, it is a continuous process. It begins in process development, continues through GMP clinical trials, scale-up and transfer to production, and then resides in the manufacturing process. Critical process parameters and critical quality attributes are identified and challenged early in development, making pre-determined process quality and failure parameters known prior to scale-up. Process validation at scale is the final integration of facility design and construction, utilities, equipment, automation, processes, procedures, and systems. The final process validation should demonstrate a consistent process outcome. The process should be understood well enough to ensure that inherent day-to-day variability does not impact process reliability. Once established, the process must be maintained and monitored appropriately throughout its life. Tools to monitor process capabilities should ultimately prove process robustness, while primarily intended to act as a trending tool for tracking and responding to negative trends before process upset.

Validation can be done retrospectively, concurrently, and prospectively. Once established, the continuing program of change control, preventative maintenance, on-going trending and training are key to continued consistent and reliable processing. The preferred validation approach will depend on the situation and the amount of risk that the company is willing to take.

Concurrent validation is accepted for certain process components, such as water systems. Water system validation is also an application where on-going end product testing is required. The PMA supports the use of concurrent validation when quality attributes can be determined with a sound sampling program and definitive and validated test methods; the testing must ensure that the desired attributes have been attained with a high level of confidence (10). Although the FDA does not directly refer to concurrent validation in their 1987 guideline, they discuss the "Acceptability of Product Testing" which parallels the concept of concurrent validation. FDA does express concerns about the inherent risks of this approach, but recognizes that the other approaches may have limited applicability in certain situations. Companies using concurrent validation must be aware of the risks involved and, in the event of an inadequate result, have contingency plans in place. An example where this approach was employed successfully is highlighted in the section on Future Trends.

Retrospective validation makes use of historical data for lots already made. Assuming a statistically significant number of lots are available to review, retrospective validation is the most accurate means of predicting the actual process capabilities. Three prospective lots can confirm the applicability of development data, but does not allow for meaningful trending. Process capability data are tracked to evaluate process variability. This can lead to critical adjustments in the process, control specifications (where appropriate), and other critical elements associated with running a process. Retrospective validation can be a useful tool for trouble-shooting, gathering data for a future product using a similar process, or re-evaluating a process that was not thoroughly validated at start-up. For example, assays associated with older in-line products may have not been validated when the product was introduced. If similar assays are to be used in a new process, they will require further testing and evaluation.

## Current trends

**Prospective Validation.** Based on many recent discussions, publications and presentations given by various scientists and regulatory experts in the industry, the current validation focus is clearly on prospective evaluations. As discussed earlier in this chapter, a sound prospective validation program requires adequate process development with a rationale long-term plan upfront. This "good science" approach can be envisioned in three phases: process characterization and optimization, process performance qualification and finally the formal process validation of the manufacturing process using three consecutive lots. Some post-market analyses can be carried out as described earlier. For instance, precise performance limits can be measured with statistical significance using the data from 25 or more consecutive manufacturing lots (11).

The first phase, process characterization, relies on several key elements. First, close cooperation between the development groups (fermentation/cell culture, recovery, purification, and formulation) is paramount to ensure a rational and consistent process definition. Miscommunication within a development team can often result in a significant loss of process control. Second, processes must be developed that are consistent, robust and, scaleable. This evaluation ideally should be made prior to the first clinical lots. Rigorous validation studies need not be performed at this point (except perhaps for viral and DNA clearance), but process controls for obvious critical variables should be defined. Changes incorporated into processes used for subsequent clinical trials and eventually commercial manufacture should be subjected to a coordinated validation evaluation assessing the full impact of the process change. Documentation explaining the rationale for the process change should be prepared.

Third, equipment must be selected that can be reliably operated, cleaned and maintained. Finally, assays must be developed of sufficient reliability, sensitivity and reproducibility to evaluate impurity levels and product concentration, activity, quality, heterogeneity, and stability. Assays need not be validated to generate development data that supports validation, but should be sufficiently developed and characterized to assure reliable data; many companies refer to this level of assay development as assay "qualification". Once these key elements are in place, the pilot batches, the transfer to manufacturing, and the final validation should be quite straight-forward.

The second phase, "performance qualification", ensures that the process performs as it was designed by establishing process control limits. Validation studies are carried out at lab or pilot scale (depending on the study) to confirm process robustness with respect to the critical parameters and critical quality attributes. This often requires using scaled-down models of the manufacturing process, which have been qualified as being representative of the full-scale process. As discussed in several chapters in this book, the use of scaled-down models is pivotal to the validation effort. Worst-case or statistical design methods are often employed to systematically challenge the process control limits.

Validation activities are generally linked closely to scale-up as the process moves closer to implementation in the manufacturing plant (1,7). Scale-up may often cause slight changes in product or impurity profiles. However, a process should be sufficiently robust to handle any changes that arise; for example, the purification steps should be able to absorb an increase in impurity level in a fermentation product that can often occur during scale-up. The activities of process qualification may have to be repeated at each scale, as scale sensitive parameters are identified. The identification of scale-related process changes can be achieved using comprehensive process monitoring to evaluate process performance during scale-up and the validation studies. When carried out properly, these studies should provide the basis of the validation program by establishing process robustness and operating limits with respect to critical parameters and scale.

The final phase of validation requires successful completion of several consecutive manufacturing lots that are properly documented following a validation protocol. In the United States, the FDA requires a minimum of 3 consecutive lots, while the European Union may require 5 lots. As discussed earlier, a few critical parameters may be varied during these lots to verify that the critical quality attributes remain within pre-defined limits. Additional process monitoring will be required across 25 or more lots to establish statistically significant process limits. Once this is done, the level of testing may be reduced.

**Coordination of a validation program and start-up**: The validation process involves a significant coordination of work across multiple disciplines and departments. A considerable amount of information must be collected, summarized and properly documented to ensure compliance. Several approaches have been recommended for accomplishing this goal (10, 12). The approach that a company may choose will depend on its size, its structure and resources. For a large company, validation involves a team effort, including a dedicated validation group. Members from various groups, including research and development, engineering/technology support, production, validation, quality control, regulatory affairs, and quality assurance, all participate in the start-up and validation activities.

The R&D group, including pilot-plant staff, develops and characterizes the processes and assists in all aspects of start-up. Together with R&D, the engineering group defines the necessary equipment and utilities, facilities and trains production and maintenance staff. This group also oversees the procurement, installation and qualification of the equipment and utilities. Production participates in the start-up and validation activities from the beginning and is responsible for writing the SOPs and the final batch records and training operations staff. The validation group coordinates the formal validation process; their primary duties include preparing the validation protocols, overseeing the validation studies (which should be carried out by production with assistance from R&D and engineering and technology support groups), processing the data and preparing the final report. The validation group should have an R&D counterpart to manage early, pre-market validation activities supporting clinical manufacturing; this will facilitate transfer of processes from the laboratories and pilot plant to the manufacturing plant. Quality control carries out the necessary tests for the validation studies as well as for product monitoring and release, while an environmental monitoring group ensures that process utilities and HVAC are fully qualified. Lastly, regulatory affairs and quality assurance groups support the effort by ensuring that the team is fully aware of all pertinent regulatory requirements from CBER and is compliant with cGMPs.

**Scope of the validation program.** Biological manufacturing is very complex and involves multiple steps, including generation of the stock and working seeds, fermentation and cell culture, recovery, purification, formulation and filling. A significant number of variables need to be evaluated to characterize and eventually validate the entire process. To evaluate process robustness with respect to these parameters, companies need to consider the critical performance parameters or attributes. These will vary depending on the product; Table I lists the most common attributes associated with biological products. This list is not all inclusive, nor does it suggest that every product be tested for all of the attributes listed. Companies need to exercise good judgment selecting the critical attributes and the corresponding assays and in-process monitoring to demonstrate process control We will see in several chapters in this book how companies have utilized statistical design methods and worst case analyses to reduce the number of parameters and thus streamline the validation process. The key is to prioritize based on criticality of the step (e.g. aseptic versus non-aseptic and final product versus intermediates) and the critical parameters and attributes.

**Importance of Communication**: The foundation of a sound validation program is open communication. The amount of inter-company communication has increased substantially; this is evidenced by the significant number of symposiums, workshops and publications over the past few years. This communication is the key to establishing standards which will streamline the validation process. Of pivotal importance is the communication within a given company. Consistent validation philosophies need to be maintained across divisions and departments. Some companies establish process validation plans early in development to determine roles and responsibilities and place boundaries on the amount of work (11, 13). Finally, companies need to communicate freely with the regulatory agencies. Since the guidelines are broadly stated, scientists and engineers need to properly interpret them for their specific processes and products. They need to use good rationale and judgment and be able to defend it to the regulatory agencies. When different procedures are used than outlined in the FDA guidelines, companies are advised to discuss the matters with the regulatory agencies to assure that they are acceptable. This can save considerable effort if deemed unacceptable at a later time.

## Future Trends

Concurrent validation studies are becoming a more important component in a validation package. These studies may employ the use of qualified (not yet rigorously validated) assays to evaluate impurity levels in various process streams. A protocol may be executed which tests the levels of impurities following product synthesis by the fermentation or cell culture, and then determines the removal of the impurities through the purification process. These data should reveal a consistent challenge to purification performance, which provides predictable impurity removal to reproducibly low levels in the final bulk drug substance. In some cases, the impurity may fall to levels below the detection limit of the assay; in these cases, a complimentary study using a scaled-down model and radioactive probes may be needed to validate impurity removal in subsequent steps. Combining the results from concurrent manufacturing-scale runs with small-scale data gathered for downstream steps can provide a compelling case for the

**Table I. Typical Quality Attributes for a Recombinant Protein Process**

| STEP | ATTRIBUTE |
|---|---|
| Seed Flask | Cell Density |
| Seed Fermentor | Cell Density<br>Carbon Evolution Rate (CER) |
| Main Fermentor | Cell Density<br>Carbon Evolution Rate (CER)<br>Oxygen Utilization Rate (OUR)<br>Product Titer/Expression level<br>Plasmid Marker Loss<br>Aerosol generation/removal |
| Cell Harvest/Removal | Solids yield |
| Cell Disruption and Debris Removal | Degree of cell lysis/% break<br>Product Yield |
| Capture Chromatography | Product Yield and Concentration<br>Purity (HPLC)<br>Product Degradation/Stability<br>Resin reuse |
| Finishing Chromatography | Product Yield and Concentration<br>Purity (SDS PAGE, Western Blots)<br>Product degradation/Stability<br>Endotoxin/DNA/ Host Cell Protein levels<br>Virus Removal<br>Levels of additives (anti-foam, etc.)<br>Resin Reuse |
| Ultra-filtration | Product Yield and Concentration<br>Purity (SDS gels)<br>Product degradation/Stability<br>Virus Removal<br>Membrane re-use/integrity |
| Viral inactivation | pH, Temperature<br>Solvent/Detergent concentrations<br>Adequacy of Mixing |
| Sterile Filtration | Product Yield/Degradation/Stability<br>Sterility |
| Formulation | Product & Excipient concentrations<br>Sterility/pH/Homogeneity |
| Lyophilization | Moisture level<br>Aggregate levels<br>Product degradation/activity |
| Intermediate Product Hold | Product Yield/Concentration<br>Product Degradation/Stability<br>Endotoxin levels/Bioburden |

elimination of routine testing for certain impurities (see the chapter by M. Leonard, et al, in this volume).

Another trend in the validation of biopharmaceutical manufacturing processes is the comprehensive evaluation of process control limits for critical processing steps; these limits shall provide adequate assurance that the process will yield product of sufficient quality when operated within these control limits. For a biological product, the complex interplay of control variables for both synthesis and purification steps requires a rigorous evaluation of these variables and their interactions, which can cause process failures. Several different approaches have been tested to this end, including fractional factorial designs, worst-case evaluations of several successive processing steps, and establishment of proven acceptable ranges. The limit of these endeavors may be difficult to define, because all process variables cannot be tested; scientific judgment must be exercised to limit the scope of the study. The evaluation of all possible sources of variation would be far too time consuming and costly.

The designation of certain products as being "Well-Characterized Biologics" from the 1995 and 1997 FDA meetings will impact process validation in a significant fashion (14, 15). Many companies producing licensed products are seeking to modify the manufacturing process to increase capacity, reduce cost, or increase product purity. The concept of product comparability serves to establish two manufacturing processes as being capable of providing comparable products. The comparability may be measured by both routine release testing and non-routine characterization testing to evaluate the physico-chemical and biological properties of the products. The removal of process-related impurities or contaminants by the modified process must also be addressed, so that an equivalent level of assurance is also established and that the new process is capable of consistently providing product of appropriate quality and safety.

The FDA allows the clinical testing of products manufactured in pilot plants prior to the construction and validation of a full-scale manufacturing plant as described in FDA's guidelines on pilot plants (16). In some cases, the lack of manufacturing data from full-scale operations may hamper the definition of appropriate in-process control limits. In these cases, applications of statistical process control have been proposed for the establishment of appropriate process control ranges (11).

## Conclusions

Process validation has finally come of age in the Biopharmaceutical industry. As illustrated in the following chapters in this book, many companies now embrace process validation as sound business and engineering practice. The concepts evolved from aseptic processing and are now being applied to bulk bio-pharmaceutical manufacturing. The authors hope that through symposiums and publications, such as this book, and continuing dialog with the FDA, a clear framework can be established for validating any bio-process.

**References.**


1. Food and Drug Administration, "Guidelines on General Principles of Process Validation,", Rockville, MD, May 1987.
2. Berry, I.R. and R. A. Nash, Ed., *Pharmaceutical Process Validation*, 2nd Ed., Marcel Dekker: New York, NY, 1993.
3. Harpaz, D. and Berry, I.R., Eds.; *Validation of Bulk Pharmaceuticals*; Interpharm Press, Inc: Buffalo Grove, Illinois, 1997.
4. Avis, K.E., In *Remington's Pharmaceutical Sciences*; Remington, J. P.; Gennaro, A.R Eds.; Mack Publishing Co.: Easton, Pa., 18$^{th}$ Ed, 1990; Chapter 84, pp. 1545-1569.
5. Lydersen, B.K., D'Elia, N.A., Nelson, K.L., Eds.*; Bioprocess Engineering: Systems, Equipment, and Facilities*; J. Wiley and Sons, Inc.: NY, NY, 1994.
6. DeSain, C., Sutton, C.V., *Pharm. Technol.*, **1995**, *19, no. 10 (October),* pp.130-136.
7. Akers, J., McEntire, J., Sofer, G. *Biopharm. Manuf.;* **1994**, *7, no. 2,* pp. 54-56.
8. Chapman, K. G., "A History of Validation in the United States- Part I", *Pharm. Technol.*, **1991**, *15, no. 10 (October),* pp.82-96.
9. Chapman, K. G., In *Pharmaceutical Process Validation;* Berry, I.R.; Nash, R. A., Eds.; Marcel Dekker: NY, NY, 1993, 2nd Ed., Chap. 20; pp. 587-597.
10. Lee, J.L., In *Pharmaceutical Process Validation;* Berry, I.R.; Nash, R. A., Eds.; Marcel Dekker: NY, NY, 1993, 2nd Ed., Chap. 19; pp. 573-586.
11. Murphy, R.; Seely, R.J., Validation of Biotechnology Bulk Pharmaceuticals, In *Validation of Bulk Pharmaceutical Chemicals*; Harpaz, D. and Berry, I.R., Eds.; Interpharm Press, Inc: Buffalo Grove, Illinois, 1997.
12. Jeater,. J. P., Cullen, L.F., Papariello, G.J, in *Pharmaceutical Process Validation*, 2nd Ed., I.R. Berry and R. A. Nash Ed., 1993, Marcel Dekker, NY, NY, p9-24.
13. Winkler, M., Presentation at the 213$^{th}$ ACS National Meeting, BIOT division, Paper No. 270, San Francisco, April 13-17, 1997.
14. FDA, 1995a, Characterization of Biotechnology Pharmaceutical Product, Dec 11-13, 1995, Washington D.C, CBER.
15. FDA, 1997, Guidance for Industry: Changes to an Approved Application for Specified Biotechnology and Specified Synthetic Biological Products, CBER and CDER.
16. FDA, 1995b, FDA Guidance Document Concerning Use of Pilot Manufacturing Facilities for the Development and Manufacture of Biological Products.

Chapter 2

# Process Characterization Studies To Facilitate Validation of a Recombinant Protein Fermentation

**Jinyou Zhang, Jay Reddy, Peter Salmon, Barry Buckland, and Randy Greasham**

**Bioprocess R&D, Merck Research Laboratories, Rahway, NJ 07065**

A fed-batch fermentation process employing chemically-defined medium was developed for the production of a recombinant protein in yeast. Process characterization studies were carried out in 23-liter fermentors to define process capability and facilitate prospective process validation at the production scale. Key to these laboratory studies was the identification of critical process parameters that determine the quality attributes desired at the fermentation stage. Among other variables, variations in inoculum age, carbohydrate feeding scheme, and harvest criterion were found to profoundly affect the fermentation performance. A thorough understanding of process capabilities was achieved and suitable working ranges for these parameters were established. With more than twenty batches carried out under the "defined" conditions, a consistent fermentation yield which met the prespecified goal was achieved. Bridging studies suggested that the pilot-scale process capability was similar to that defined in small-scale process characterization studies, thus setting the stage for rapid process qualification and process validation at production scale.

The final phase of introducing a new fermentation process to manufacturing is usually its validation, or "to establish documented evidence that provides a high degree of assurance that a specific process will consistently produce a product meeting its pre-determined specifications and quality attributes" *(1)*. Besides operational qualification and performance qualification of the equipment, this validation activity includes defining the critical quality attributes (e.g. product yield, cell density, culture purity, etc.), and determining the operational ranges of the critical process parameters (e.g. inoculum state, nutrient feed scheme, harvest time, etc. in the process described below) within which the process will consistently achieve the critical quality attributes. However, determining the operational ranges at manufacturing scale can be quite laborious, time consuming and costly. A practical alternative approach is to perform these "range" or characterization studies in laboratory fermentors and demonstrate that they are insensitive to scale-up. These operational ranges can then become part of the validation package and be confirmed during the manufacture of the consistent lots. Only the critical process parameters that are found to be sensitive to scale-up would require "range" studies at manufacturing scale.

This paper describes the characterization studies in laboratory fermentors for a recombinant yeast fermentation process producing a therapeutic protein, and the bridging studies up to pilot scale.

## Materials and Methods

**Cultures and Chemicals.** The culture used in this study was a yeast recombinant strain which contains genes coding for the therapeutic protein under regulatory control of the *GAL*10 promoter. The genotype of the host strain is *MATa leu2*-04 *mmn9::URA3*, *ade1* cir$^{o}$. The working stocks expanded from a master frozen vial were stored at -70°C as frozen bottles (150 ml/250-mL Nalge bottle) in the presence of 12.5% glycerol.

All chemicals for medium make-up and analysis were purchased from Sigma Chemical Co., EM Science and Fisher Scientific. Assay kits for the protein product were bought from Abbott Laboratories.

**Culture conditions.** For inoculum development, a leucine-free defined medium as that described by Fu *et al.* *(2)* was used for all stages of inoculum development. The number of stages and inoculation volumes were the scaled-down versions of the prospective manufacturing process. One hundred and fifty milliliters of frozen seed (contained in a 250-mL bottle) were thawed and used to inoculate a 23-L fermentor (D1) containing 12 liters of the medium. The culture was cultivated at 28°C, 600 rpm of agitation and 6 L air/min until carbon evolution rate (CER) dropped which indicated depletion of glucose, and was transferred at 7.5% (v/v) to another 23-L fermentor (D2) operated under the same conditions. Immediately after the CER dropped, the culture was inoculated to a production fermentor at 8% (v/v). Except for the difference in scale, the pilot scale inoculum development was the same as that of the lab scale described here.

For production fermentation, a chemically-defined medium containing no leucine was used for all production fermentations. The medium was developed and optimized based on that of Oura *(3)*, with modification of some components and addition of specific growth factors. A fed-batch operation was employed with glucose feed to support biomass build-up, followed by galactose feed for protein production. The lab-scale fermentations were carried out in 23-L fermentors containing 15 liters of medium. The tanks were operated at 28°C with an agitation of 600 rpm, an aeration of 12 L/min (0.8 vvm), and a back pressure of 0.6 bar. Respiratory activities such as OUR (oxygen uptake rate) and CER, dissolved oxygen (DO), and pH were monitored on-line. The 1000-L pilot scale fermentations were carried out using the same process, with the operational parameters adjusted accordingly. Specifically, the aeration was started at 0.5 vvm and the agitation at 100 rpm; during peak oxygen demand period, the aeration was increased to 1.0 vvm and the agitation to 300 rpm.

**Analytical Methods.** Growth was determined by measurement of dry cell weight (DCW). Carbohydrates and alcohols including glucose, galactose, glycerol and ethanol were analyzed by an HPLC system off-line using broth supernatants. To profile the protein production, cell pellets containing ~ 20 mg DCW were prepared from fermentation broth samples taken at various time points, washed once with PBS buffer and stored at -70°C till breakage. The cells were broken by vortexing in the presence of glass beads and the protein product concentration was determined by an enzyme immunoassay (EIA). All results were back-calculated and expressed as fermentation titers (unit/L).

Due to the inherent variation of the EIA used, the comparisons listed in each table or figure were based on the assays carried out at the same time and under the same conditions for the experimental runs and the respective controls to minimize variations due to assay kits, standards, and assay conditions. Similarly, all the comparisons were based on the fermentations carried out at the same time or under similar conditions to eliminate culture differences. When more than one fermentation or measurement were carried out, average results were reported.

## Results and Discussion

The fermentation process we developed is shown in Figure 1. A large size of frozen seed (150 ml) is thawed and transferred directly to a stirred-vessel for cultivation. The culture is inoculated to a second stage seed tank and then to the production tank based on the on-line respiration profiles (CER). The production stage employs a chemically-defined medium and a two-step feed approach. Glucose feed builds up cell mass and galactose feed follows to induce and supply energy for the protein product synthesis. The initiation of glucose feed and the determination of harvest time are also dependent on the on-line CER profiles. Kinetics from a typical fermentation are given in Figure 2. Process characterization studies were carried out in 23-L laboratory fermentors in order to define the process capabilities and to facilitate the prospective process validation to be performed at manufacturing scale. Due to the direct impact of harvest timing in the evaluation of all other process parameters, determination of harvest window is described first below.

**Window for harvest**. To ensure consistent quality of fermentation broth, the initiation of cell harvest should depend on a physiological event of the culture rather than time. The harvest criterion was established as the precipitous drop of CER which indicates the depletion of galactose and thus the discontinuation of protein synthesis as shown in Figure 2. It can be seen by the example in this figure that

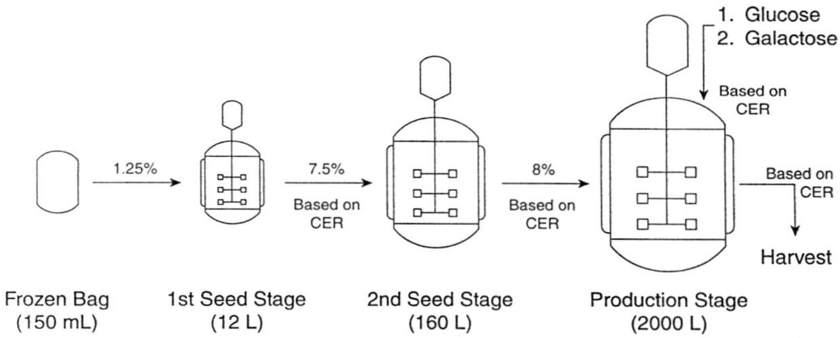


Figure 1. Fed-batch defined medium fermentation process for production of a therapeutic protein by a recombinant yeast

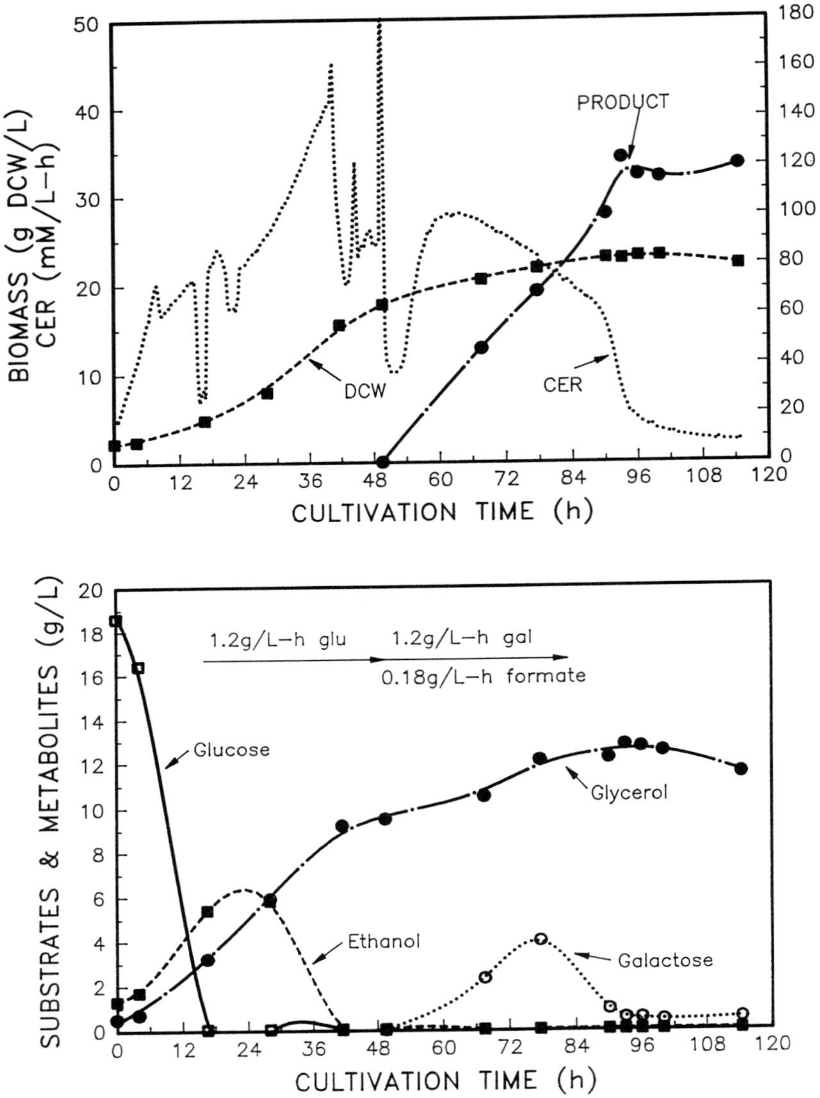


Figure 2. Determination of harvest window based on on-line respiration profile in 23-L fermentor

when the fermentation was extended by 10~20 hours after the CER had dropped, no significant change in biomass or product titer occurred. Since there was no further synthesis of the protein product due to the depletion of inducer, and since the product titer was not changing (within assay variation), product degradation was essentially negligible. The "viable counts" based on CFU (colony forming unit) on plates remained constant just before and after CER drop (88 ~ 116 hours) (not shown). It was therefore concluded that the window for cell harvest can be practically set within 10 hours after the CER precipitously droped. Initiating harvest after the drop is critical, because prior to this event, there is no clear and distinctive on-line characteristic and the titer is rapidly increasing, making the delivery of reproducible and high-yielding broths practically not achievable.

**Inoculum transfer criteria**. The fermentation process employs seed cultures that are transferred at rapidly dropping CER ("falling CER") which indicates the depletion of glucose. But according to conventional wisdom, the inoculum should be transferred before glucose is depleted and when the culture is still in the exponential phase in order to minimize growth lag *(4)*. "Early" inoculation was studied by transferring the culture at peak CER, at which time there was still 10 ~ 15 g/L glucose remaining. As shown in Figure 3, "early" transfer led to only gradual instead of the typical precipitous drop of CER at harvest stage, apparently due to the inefficient utilization of galactose by the culture at production phase (data not shown). Presumably, the inoculum was still at the "uninduced" state in terms of oxidative utilization of carbon source. As a result, the distinctive harvest criterion disappeared. While the slower utilization of galactose due to "early" inoculation did not affect cell mass, it did extend the presence of the inducer during the product synthesis phase, which resulted in increasing fermentation titer as the harvest time extended (not shown). For the process validation purpose of this biologic fermentation, however, consistency rather than yield improvement was the primary target. Since no definitive on-line harvest criterion could be followed with "early" inoculation, inconsistent broth may be delivered to downstream processing.

On the other hand, "late" inoculation would also change the production fermentation profiles. It was observed that when the inoculum was transferred several hours after CER had reached baseline, the initial glucose utilization became less efficient during production fermentation and the utilization rate was decreased considerably which led to a slower growth rate. As can be seen from Figure 3, the batch with "late" inoculation did not deplete glucose at the normally seen ~16 hrs because no CER drop was observed (at which time glucose feed was still started in this batch), resulting in a delay in galactose feed initiation and thus delay in broth harvest. In conclusion, subtle changes in inoculum physiology will result in dramatic differences in production fermentation characteristics, and a proper window for transfer should be within about 2 hours from CER falling to half of its maximum value.

**Glucose feed rate and amount**. Glucose feed is critical to this fermentation in that it builds up $\geq$ 80% of the biomass, the machinery for product synthesis. The feed was designed to be at a limiting rate so that the Crabtree effect (repression of

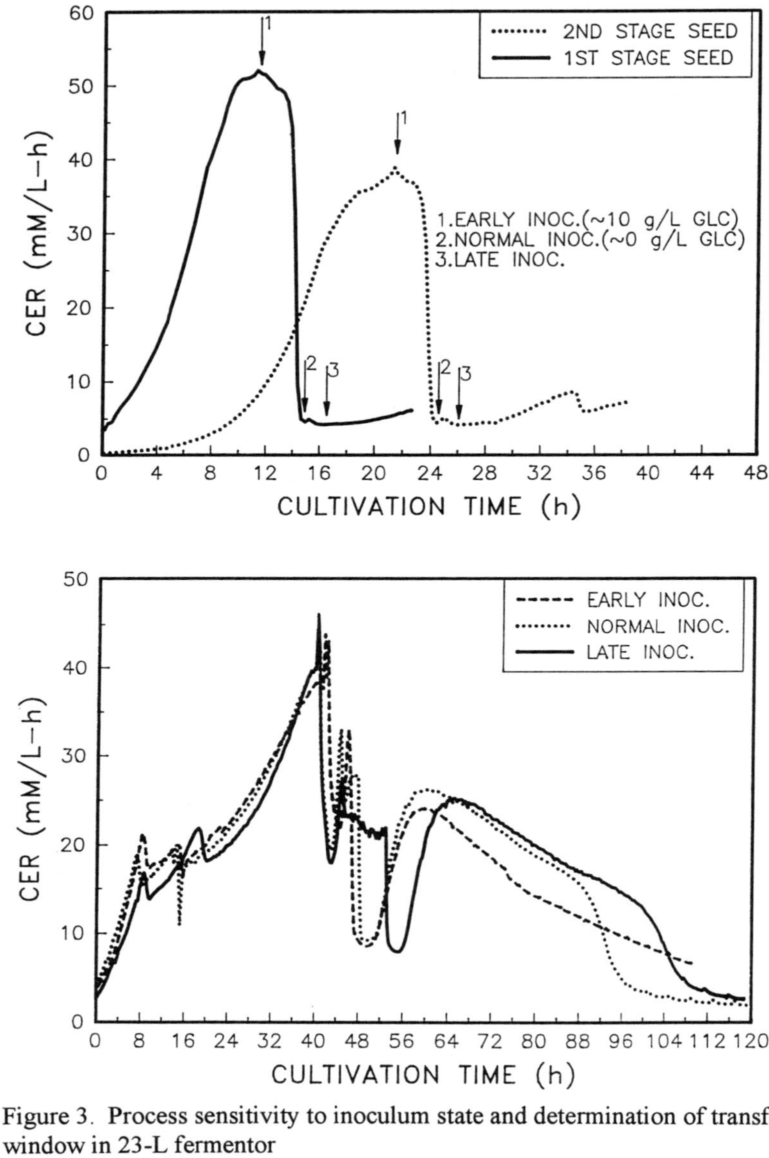


Figure 3. Process sensitivity to inoculum state and determination of transfer window in 23-L fermentor

oxidative metabolism by excessive glucose concentration) could be avoided. The effect of various feed rates and feed amount was studied in order to determine the ranges within which the process outcome would consistently meet the prespecified target. Figure 4 describes the results of studies on glucose feed rate. Under the regular glucose feed amount of 40 g/L, varying the feed rate by ± 0.2 g/L-h from the regular 1.2 g/L-h led to a significant decrease in product titer at both ends, suggesting a narrower window. While 1.3 g/L-h exerted no effect, 1.1 g/L-h still delayed and lowered product synthesis compared to the 1.2 g/L-h batch concurrently run and assayed (Figure 4). The acceptable range appeares to be 1.25 ± 0.10 g/L-h.

Figure 5 shows the effect of various glucose feed amount on the fermentation performance. Under the same glucose feed rate of 1.2 g/L-h and with galactose feed started only after completion of glucose feed, feeding 50% more glucose (total 60 g/L) than the regular 40 g/L resulted in a dramatic reduction of product yield. Glucose variations within only 10% above 40 g/L did not appear to have a significant impact (Figure 5). Since cell mass production is directly linked to the amount of glucose feed, a feed lower than 40 g/L will result in lower cell mass for product synthesis, and is therefore not desirable.

**Galactose feed rate and amount**. Galactose is the carbon/energy source as well as the inducer for protein product synthesis, and its variation is likely to affect the fermentation. Compared to the regular feed amount of 40 g/L, feeding 25% more galactose (total 50 g/L) expectedly led to about 10-h delay of the precipitous CER drop, the harvest criterion (Figure 6). While biomass production was not sensitive to this variation because it was mainly built up by glucose feed (Figure 2), the availability of more inducer and energy source led to ~30% higher production after 110 hours (Figure 6). In a further study, increasing the feed to 50% more (total 60 g/L) still led to only ~30% higher titer than that by the control (Figure 6). Therefore, the important fermentation quality attribute or the product yield is sensitive to galactose feed amount and higher production is achievable by extending galactose feed, but not without a limit. For reproducibility of the process, galactose feed amount should be tightly controlled at a level between 40 and 50 g/L.

Galactose utilization rate by the culture was considerably slower than the regular feed rate of 1.2 g/L-h (Fig. 2). In fact, galactose accumulation was observed irrespective of the change in feed rate from 1.0 to 1.4 g/L-h. Therefore, at a constant galactose feed amount of 40 g/L, no changes in product titer and cell density were observed under various feed rates. It is concluded that the galactose feed rate is not a critical process parameter that impacts the fermentation performance.

**Critical process parameters and quality attributes**. Among the many process parameters which may have certain impact on the fermentation performance, some were shown to be critical based on strong technical rationale, because their variations profoundly affected the important fermentation quality attributes, including the productivity and its reproducibility for the protein product and the biomass. As discussed above, inoculum transfer age, carbohydrate feed and harvest timing are apparently critical parameters, while DO level (see below) and medium sterilization are examples of noncritical ones as far as this process is concerned. It was found

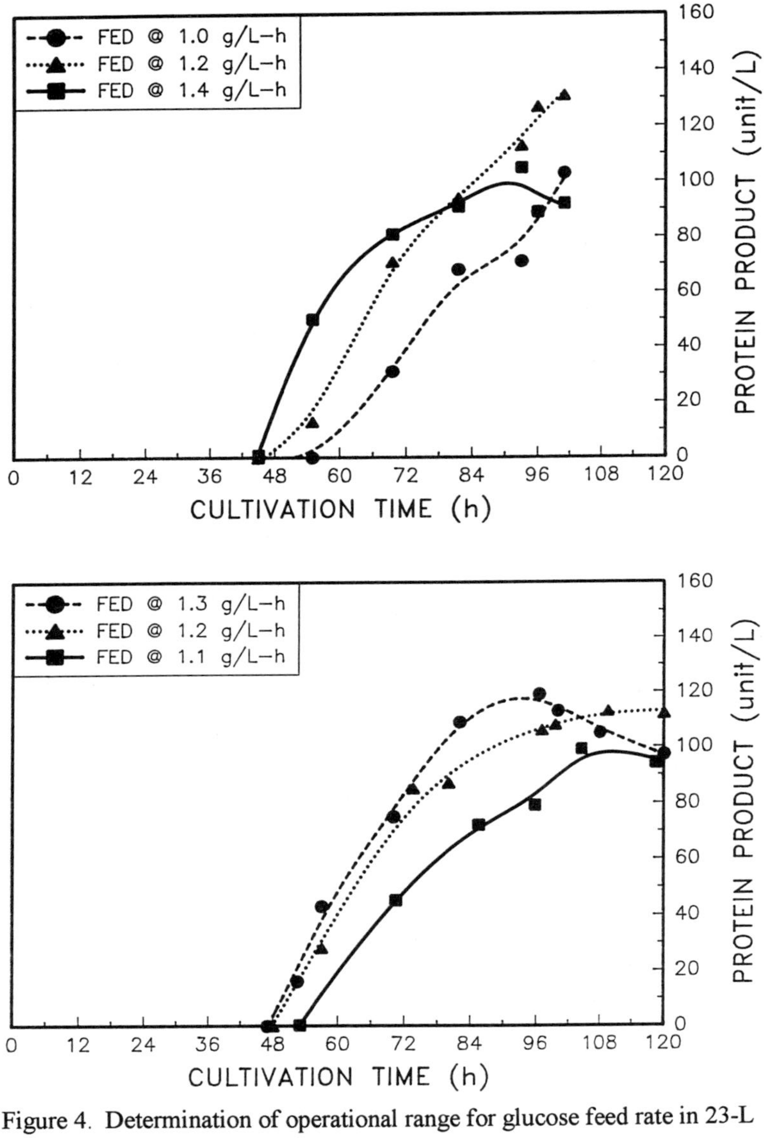


Figure 4. Determination of operational range for glucose feed rate in 23-L fermentor

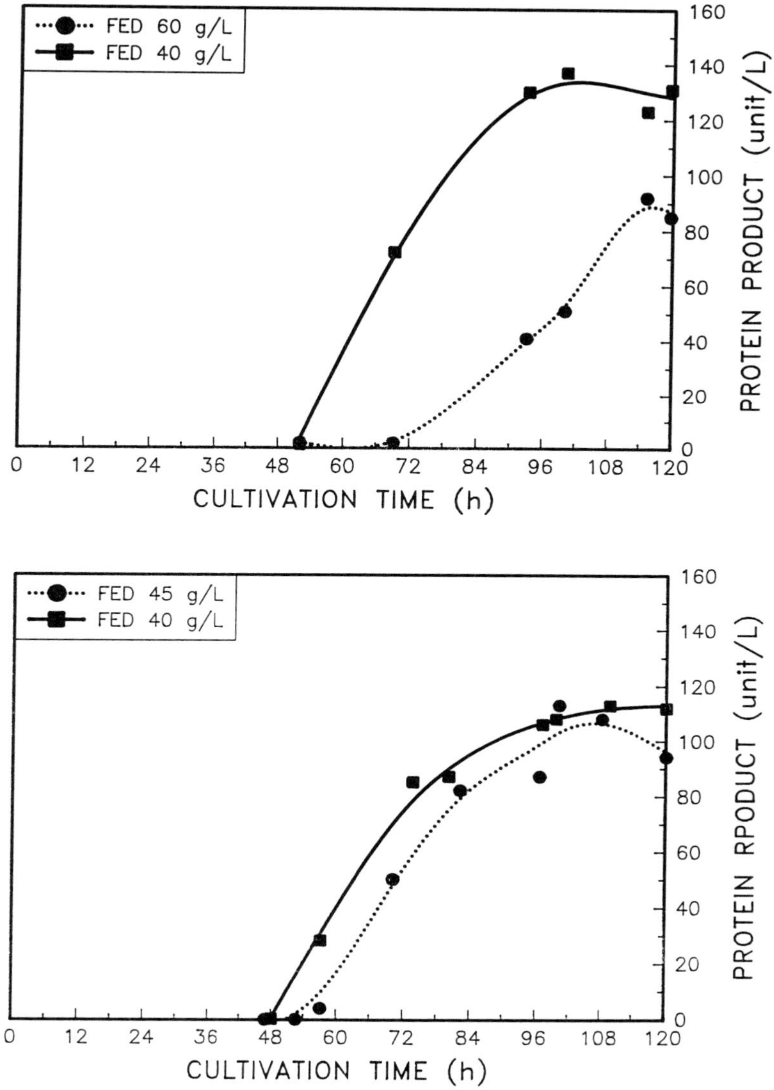


Figure 5. Determination of operational range for glucose feed amount in 23-L fermentor

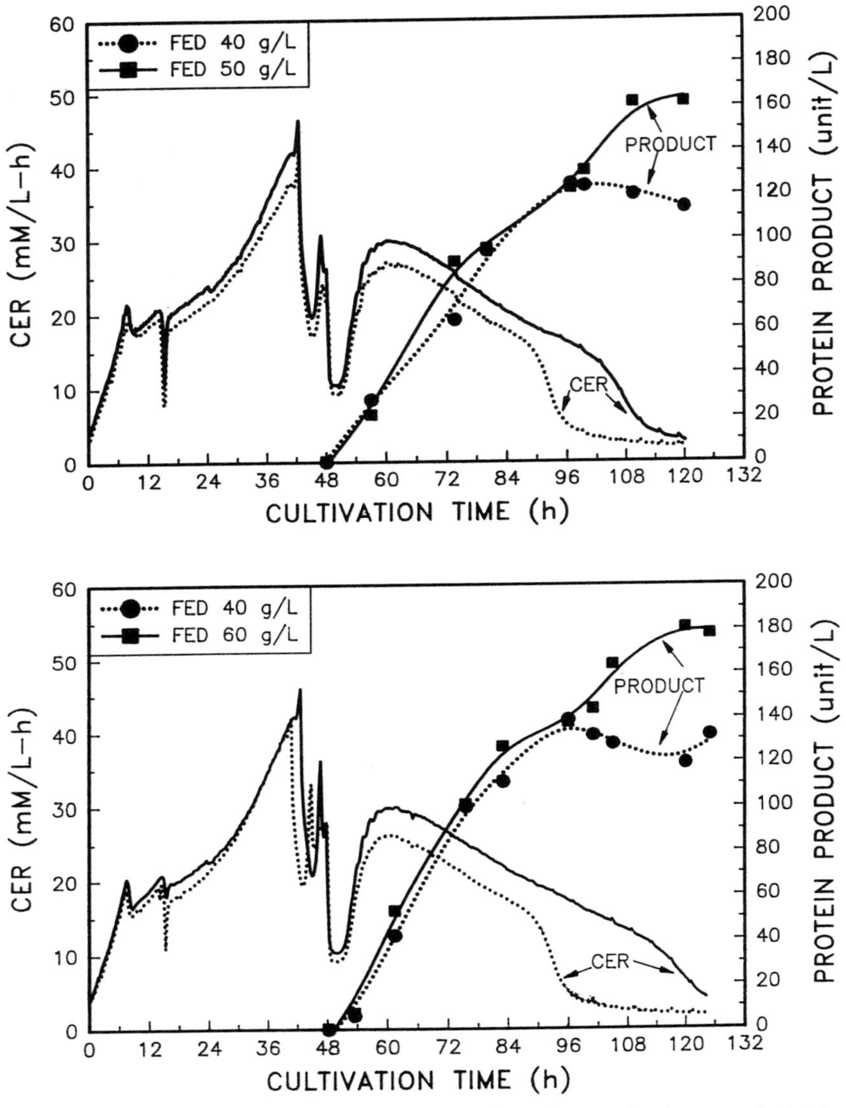


Figure 6. Determination of operational range for galactose feed amount in 23-L fermentor

that doubling the sterilization time did not change fermentation outcome at all ( data not shown), which is not surprising for a chemically-defined medium fermentation.

**Bridging between small and large scales**. All the above studies were performed at the 23-L lab-scale. Although ideal, it is not practical to carry out all of these studies at manufacturing scale. The conclusions obtained from the small scale must therefore be shown through bridging studies to hold for large scale fermentations. Figure 7 compares the respiration profiles from 23-L lab- and 1000-L pilot-scales. At both inoculum development and production fermentation stages, the CER profiles were very similar for the two scales, which indicates the similar carbon source utilization profiles and growth kinetics. Although not given here, the product titers from the two scales were expectedly found quite comparable based on the specific assays for the protein product. These results suggested the scalability of the small scale results to large scale process. Table I shows that at the 23-L lab-scale, when the fermentations were carried out within the specified windows for the critical parameters (i.e. under the"characterized" conditions), the process yielded results that were consistent and that met the predetermined specifications. Figure 8 demonstrates the extreme consistency of the fermentation process at 1000-L pilot-scale carried out under similar conditions, represented by the on-line OUR profiles. Note that the batches shown were run as far as six months apart, yet the kinetics at both growth and production stages were essentially superimposable, suggesting the robustness of this "characterized" process. Although more bridging studies are needed, conclusions from the 23-L characterization studies appear applicable to large scale process because of the lack of sensitivity these critical parameters have to scale-up, thus setting the stage for the process validation at the manufacturing scale.

Table I. Productivities by 23-L batches using parameters within specified windows

| | Total Number (n) | Biomass (g DCW/L) | Protein Product (unit/L) |
|---|---|---|---|
| Fermentation Batches | 25 | 23 ± 0.9 | 126 ± 16 |
| EIA Assay Internal Control* | 35 | N/A | 119 ± 22 |

* All internal controls were prepared at the same time from the 96-h broth of a single batch and stored at -70°C until use. The standard deviation in the results of internal control suggest the inherent variation associated with this EIA assay. N/A: not applicable.

Differences do exist in some other profiles, with the most significant one to be the DO level, in that the DO at the 1000-L scale fell to much lower level than that at the 23-L scale due to the lower power input per unit volume in the 1000-L fermentor (starting at about 25% of that for the 23-L fermentor) (Figure 9). In order

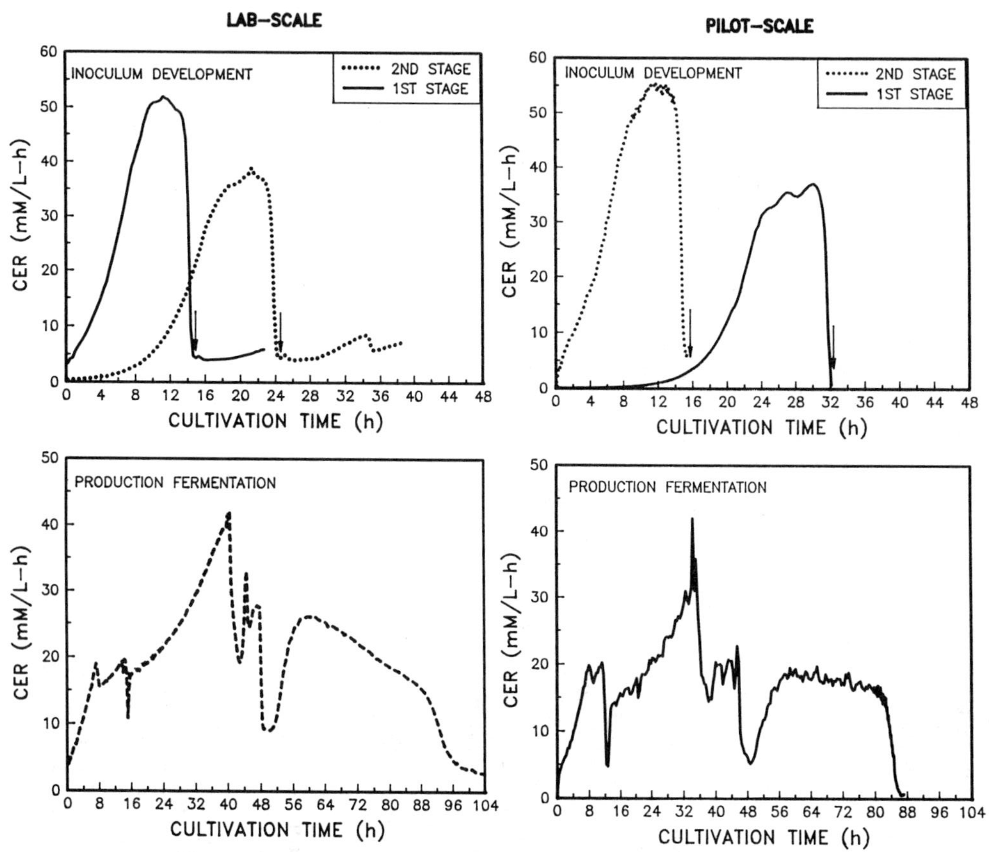


Figure 7. Bridging between laboratory (23-L) and pilot-scale (1000-L) fermentations: comparison of respiration profiles

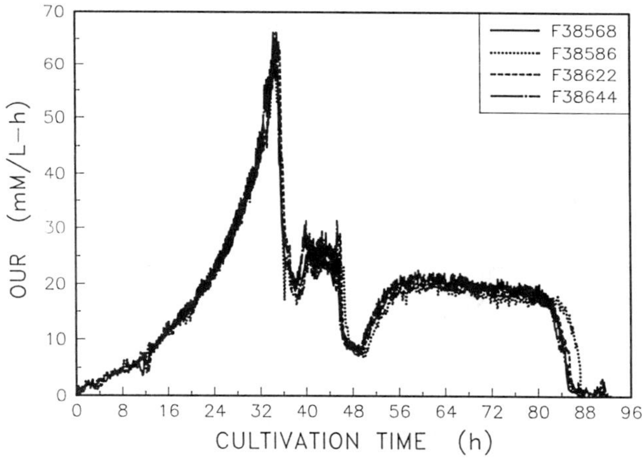


Figure 8. Consistency of 1000-L pilot-scale fermentations under the characterized process conditions

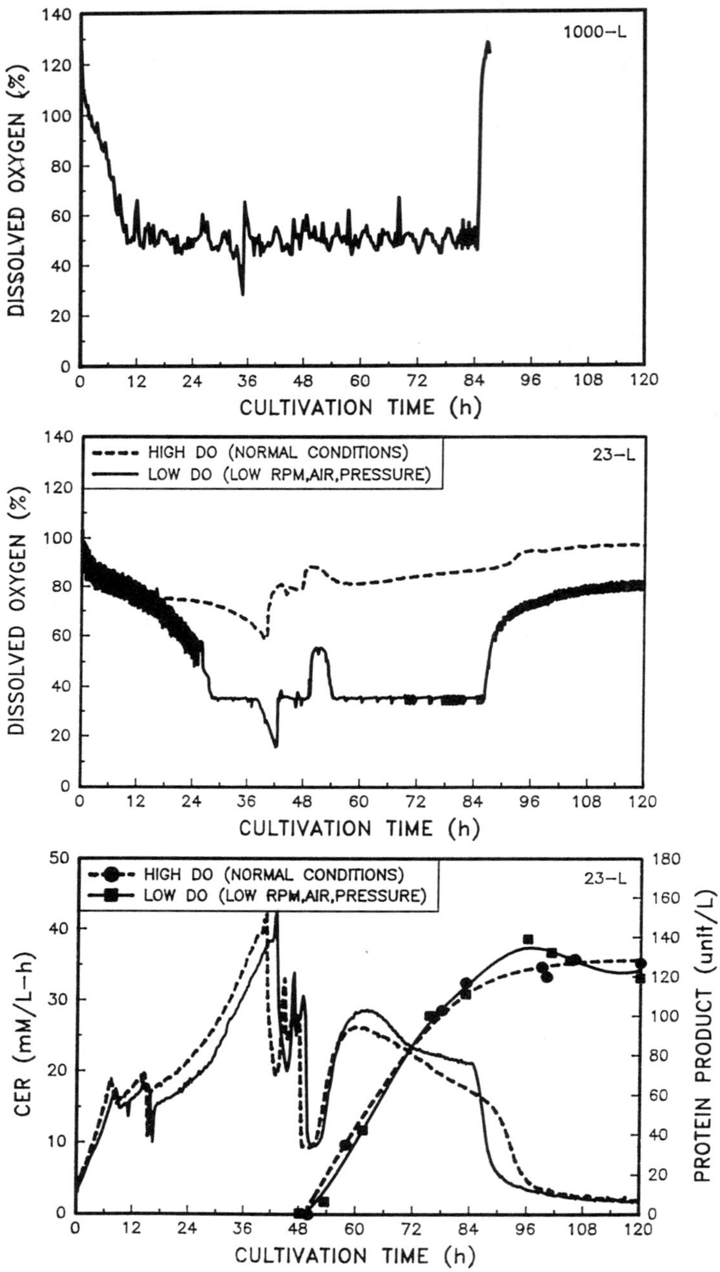


Figure 9. Bridging between laboratory and pilot-scale fermentations: noncritical difference in dissolved oxygen level

to study the potential effect of this DO difference, the 23-L conditions were adjusted by decreasing agitation, aeration and back pressure to create a low DO situation (~35% of saturation) similar to that observed at the 1000-L scale. It can be concluded from Figure 9 that there was essentially no impact on fermentation productivity by the significant DO difference experienced at lab- and pilot-scales.

## Summary

During fermentation development at laboratory scale, the quality attributes that can be achieved by the process were established. To consistently achieve these quality attributes at production scale, each of the critical process parameters (those with very narrow operating ranges) must be identified and their operational ranges defined through process validation. As demonstrated from our studies, the use of laboratory fermentors to characterize the critical and differentiate the noncritical fermentation process parameters appears to be a practical and effective tool to facilitate the ultimate validation at manufacturing scale. By operating within the ranges established for the critical parameters (inoculum state, carbohydrate feeding, and harvest timing), process consistency was clearly shown at both laboratory and pilot scales, suggesting their insensitivity to scale-up. It is therefore likely that the ranges established by the laboratory scale studies are applicable at manufacturing scale. Parameters such as galactose feed rate, dissolved oxygen level and sterilization intensity are important but found noncritical, allowing their operational ranges to be quite broad. Although less likely, further characterization studies at laboratory scale may be required once production of consistency lots commences.

## Acknowledgments

The authors would like to thank Bruce Burgess for batching all 23-L fermentors, Tom Brix for carrying out 1000-L fermentations, and other members in Bioprocess R&D for their contributions to this project.

## Literature Cited

1. FDA. Guidelines of general principles of process validation. **1987.**

2. Fu, J.; Vandusen, W. J.; Kolodin, D. G.; O'Keefe, D. O.; Herber, W. K.; George, H. A. Continuous culture study of the expression of hepatitis B surface antigen and its self-assembly into virus-like particles in *Saccharomyces cerevisiae*. *Biotechnol. Bioeng.* **1996**, *49*: 578-586.

3. Oura, E. Effect of aeration intensity on the biochemical composition of baker's yeast. I. Factors affecting the type of metabolism. *Biotechnol. Bioeng.* **1974**, *16*: 1197-1212.

4. Pirt, S. J. *Principles of microbe and cell cultivation*. Blackwell Scientific Publications, Inc.: Palo Alto, CA, 1975; Chapter 19, pp 196-198.

Chapter 3

# Validation of Continuously Perfused Cell Culture Processes for Production of Monoclonal Antibodies

**S. C. Whitaker[1], R. Francis[2], and R.C. Siegel[1]**

**[1]Centocor, 200 Great Valley Parkway, Malvern, PA 19355**
**[2]Centocor B. V., P.O. Box 251, NL-2300 AG Leiden, Netherlands**

Process validation of a continuously perfused mammalian cell culture process for the production of therapeutic monoclonal antibodies was performed, and included an analysis of the cell culture process, the production and characterization of cell banks, and the properties of the cells and product during both the growth and production phases of the cultures. The cell culture process was designed to be a closed system to prevent cross contamination. The cell banks were analyzed and shown to produce a constant amount of antibody, and to be stable. During a 35 day perfusion process the cells may undergo as many as 51 doublings and the growth rate may vary from 20 hours during the growth phase to up to 240 hours during the production phase. High viability and specific productivity were controlled and documented throughout the perfusion process. Final product was produced over a long period of time, therefore, studies were also performed to demonstrate that the product characteristics did not vary significantly during the harvest period.

We have developed a continuous perfusion bioreactor process for production of monoclonal antibodies. This process incorporates a 50 L perfusion bioreactor to generate sufficient cells to inoculate a 500 L production perfusion bioreactor (see Figure 1). The cell culture is maintained at high cell density (HCD) by means of a spin filter retention device (*1-4*). Fresh media is continually perfused through the bioreactor, and product harvested for periods of 1 to 4 months, during which the environmental conditions and nutrient concentrations remain constant. (*5,6*). Maintaining long term cultures with high viability is quite different (*7*) from traditional large scale batch cell culture practices in which cells grow for a short period of time after which the culture is allowed to die (*8-10*). Most large scale suspension culture processes are operated as batch or fed batch processes, where the environment is continuously changing as nutrients are depleted and metabolites and product accumulate (*11,12*). There are three main areas which are key in validating perfused cell culture manufacturing processes; 1) cells must be shown to be genetically stable, 2) the process must demonstrate consistent operation, and 3) product characterization must demonstrate consistency throughout all stages of the production process.

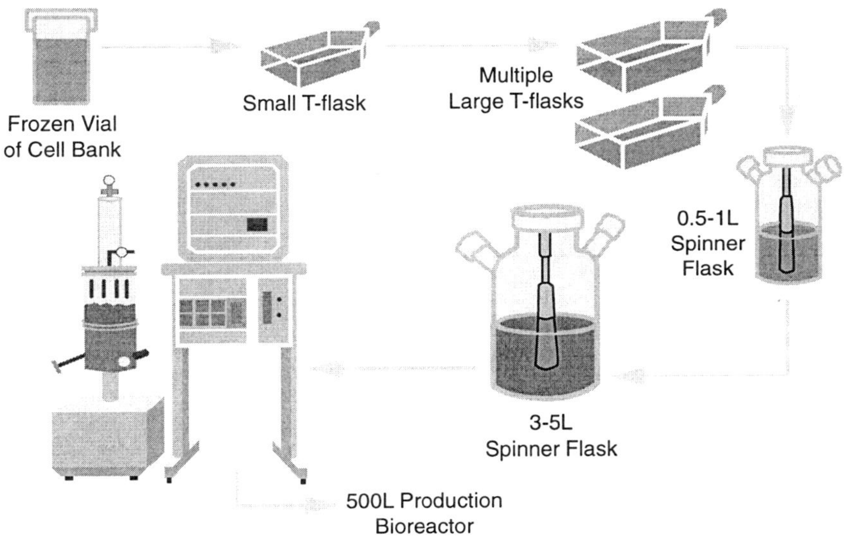


**Figure 1. Process Overview**

## Material and Methods

**Cell Lines and antibodies.** The cell lines described in this study were two murine/human IgG1 chimeric antibody producing cell lines developed at Centocor Inc., (cell lines C116, anti human IIb/IIIa and, C168, anti human TNF $\alpha$). The cell lines were created using the murine myeloma Sp2/0-Ag14 host cell. Details of the origin and development of the cell lines are described elsewhere (*13,14*).

**Culture Media**. The serum containing medium used for cultivating C116 and C168 cells (FBS) comprised Iscove's Modified Dulbecco's Modified Eagle's Medium (JRH Biosciences, Lenexa, Kansas, USA) supplemented with 5% (v/v) heat inactivated fetal bovine serum (JRH Biosciences, or, Intergen, Purchase, New York, USA), and Pluronic F68 (BASF Corporation, Wyandotte, Michigan, USA). At high cell density in the bioreactor, the FBS medium also contained a protein hydrolysate (Quest) at a concentration of 2.5 g/L. The serum free medium (SF) used for cultivating the C168 cell line only comprised a modified Iscove's basal medium supplemented with amino acids, vitamins, bovine insulin and bovine transferrin (JRH Biosciences) bovine serum albumin (BSA; Intergen) lipids, hormones, and Pluronic F68 (BASF Corporation) and 0.5 mg/L mycophenolic acid, 2.5 mg/L hypoxanthine, and 50 mg/L xanthine (MHX [Sigma Chemical Company, St. Louis, Missouri, USA]). Both the FBS and SF media were filter sterilized through 0.1 μ filters (Millipore, Marlborough, Massachusetts, USA) and stored at 2-14°C prior to use.

**Stability Study Procedure.** Cells were maintained in logarithmic culture by passaging every two to four days throughout a period of 12 weeks (about 100 generations). Cell density and viability were determined by staining the cells with 0.08% trypan blue in normal saline and then counting them with a hemocytometer. An average of five squares or at least 200 cells was used to determine the viable cell count and mean doubling time (MDT). Doubling times were calculated by the equations:

$$D = \ln 2 \; (C_f / C_o) \qquad (1)$$

$$DT = t/D \qquad (2)$$

Where D = number of doubling, $C_f$ = final cell concentration, $C_o$ = initial cell concentration , DT = doubling time, and t = incubation time.

After counting, the cells were subcultured to a seeding density of $1\text{-}3\times10^5$/mL in new 75 $cm^2$ T-flasks or spinner flasks (250 mL working volume), using SF or FBS medium. Each week for the duration of the 12 week study, additional 25 $cm^2$ T-flasks ($n \geq 2$) were inoculated at $1\times10^5$ cells/mL and allowed to overgrow without subculturing for 7 days. At the end of the 7 day period, the overgrown culture was sampled for a cell count and antibody analysis was performed by rate nephelometry. In this way antibody levels were constantly monitored for the duration of the stability study period. At the end of the 12 week period a small bank of frozen viable cells was secured and the cells subsequently screened to determine the fraction of cells which were producers.

**Clonality.** Cells were cloned by limiting dilution in 96 well plates and incubated for 10-14 days. Supernatant was harvested from those wells derived from a single clone and an ELISA was performed to specifically detect the presence of human IgG1. The percent producers was calculated by dividing the number of IgG1 positive wells by the total number of single clone-containing wells.

**Continuous Perfusion Process**. 500 L continuously perfused bioreactors (MBR, Switzerland) fitted with a rotating spin filter device were used for continuous large scale production; the mode of operation has been described in detail elsewhere (*3,15*). The bioreactors were operated as closed systems to maintain sterility.

**Cell Banking.** A two tier seed lot system was used to ensure uniform and consistent production of antibody (*16,17*). Each Manufacturer's Working Cell Bank (MWCB) was derived from a single Master Cell Bank (MCB) vial and each bioreactor inoculum from a single MWCB vial. After production, End of Production Cells (EPC) and Late Extended Cell Banks (LECB) were established. Approximately 25 ampules of cells were frozen as the EPC directly from the bioreactor with the addition of 20% (v/v) FBS and a cryopreservative, 10% (v/v) dimethyl sulfoxide (DMSO). The LECB was cultured in T-flasks or spinners for an additional length of time equal to 1/3 of the total production run time. A similar number of ampules containing the LECB were also cryopreserved. These banks were analyzed and compared to the MCB and MWCB to confirm stability.

**Antibody Concentration.** Cell culture samples were centrifuged to remove the cells. Antibody concentration was measured by rate nephelometry (Array Protein System, Beckman, USA). Anti-human IgG reagent (Beckman, USA) was mixed together with sample in the presence of PEG Buffer (Beckman). Rate units were measured against 10$\mu$g/mL to 20 $\mu$g/mL of purified human murine chimeric IgG1 antibody to create a standard curve. The values of unknowns were determined by interpolation from the standard curve.

**Antibody Immunoreactivity.** The immunoreactivity of the anti IIb/IIIa antigen-binding fragment (Fab) was measured by ELISA. The Fab, bound to the solid phase was detected by reaction with a second antibody, alkaline phosphatase-conjugated goat anti-human IgG (heavy and light chain), and its subsequent hydrolysis of para-nitrophenylphosphate. The colored product produced by the reaction was quantified spectrophotometrically by absorbance at 405 nm. From the Reference Standard, a binding curve was constructed by plotting the natural logarithm of the concentration against the natural logarithm of the $OD_{405,}$ divided by the $OD_{405}$ at 8 ng/mL of the Reference Standard [Ln ($OD_{405}$/$OD_{405}$ 8 ng/mL Reference Standard)] and performing linear regression. A binding curve was constructed for the test sample. The antilog of the difference in x-intercept between Reference Standard and sample is a measure of the immunoreactivity of the sample (relative to the Reference Standard).

**Antibody Structure.** Structural analyses included tryptic peptide maps, and molecular weight by mass spectroscopy. Tryptic digests were performed by reducing the intact monoclonal antibody in 4 M urea with 5 mM DTT for 15 minutes at 50°C. The reduced antibody was then alkylated with a 2-fold excess of iodoacetamide. The reduced and alkylated preparation (about 100 µg) was then digested with 2 µg of trypsin for 24 hours at 37°C. The peptide fragments were resolved by reverse phase HPLC using a VYDAK C18 column (2.1 x 250 mm) with a acetonitrile/trifluoracetic acid mobile phase and linear gradient elution.

Direct molecular weight determination of the intact C116 Fab was measured by electrospray ionization mass spectroscopy using a Sciex API 1 LS/MS System (Taylor Technology, Inc., Princeton, New Jersey, USA). The instrument was tuned and calibrated to within 0.1 amu with a 0.1 mM polypropylene glycol standard. Myoglobin was also analyzed daily to verify the calibration of the instrument. Samples were prepared in 5% acetic acid and injected into the instrument at 2.5 µL/min. In between samples the syringe, inlet tube and atmospheric pressure ionization (API) source were flushed with 5% acetic acid to prevent sample cross-contamination.

## Results

**Cell Bank Stability Studies**. Figure 2 shows data from three cell bank stability studies. Figure 2A shows the results of a stability study on an early C116 bank. Antibody levels increased for about 5 weeks then show a steep decline in productivity. These cells appear to reach a stable state after 8 weeks, but the level of antibody had decreased to 10 µg/mL or less. This was considered an early or "first generation" cell bank. Since this cell line and media were not fully optimized, the antibody (Ab) concentrations were low and there was a large amount of scatter in the data. An MCB was subsequently established by subcloning and selection of high producers. This "second generation" cell bank was shown to be stable with respect to continuous chimeric antibody production for 11 weeks (Figure 2B). A second cell line, C168 shown in Figure 2C, was also shown to be stable over a period of 12 weeks. During this 12 week period approximately 100 generations accumulated for all 3 banks.

**Accumulation of Generations**. Table 1 shows the average generations accumulated during different stages of the production process from four representative cultures of C116. Cells grow rapidly during preculture with a mean doubling time (MDT) of about 20 hours. MDT increased during later stages of production as the cells reach and maintain high cell density, giving rise to correspondingly fewer generations. The cumulative generations during each of these production runs was 22.8 to 39.7; well within the demonstrated stability of the MCB and MWCB cell banks (about 100 generations).

**Bioreactor Validation**. The process validation program for the chimeric IgG1 producing cell line C116 employed 14 bioreactors. Figure 3 shows the cell growth of 5 of the 500 L production cell cultures performed over a one year period during the validation campaign. The data showing cell growth from thawing the MWCB until HCD is reached, are plotted as relative cell number versus days. The expansion of preculture was cut back at various time points during the production process (reflecting in decline in total viable cells) on days 11, 12, 22 and 25 to maintain a cell density at a predetermined level while waiting for preparation of the inoculum bioreactors. Overall, the parallel growth curves demonstrate a very consistent cell growth profile during preculture, inoculation and growth to HCD. Once at HCD, the cultures were maintained at a constant density throughout the production phase until day 35 post bioreactor inoculation.

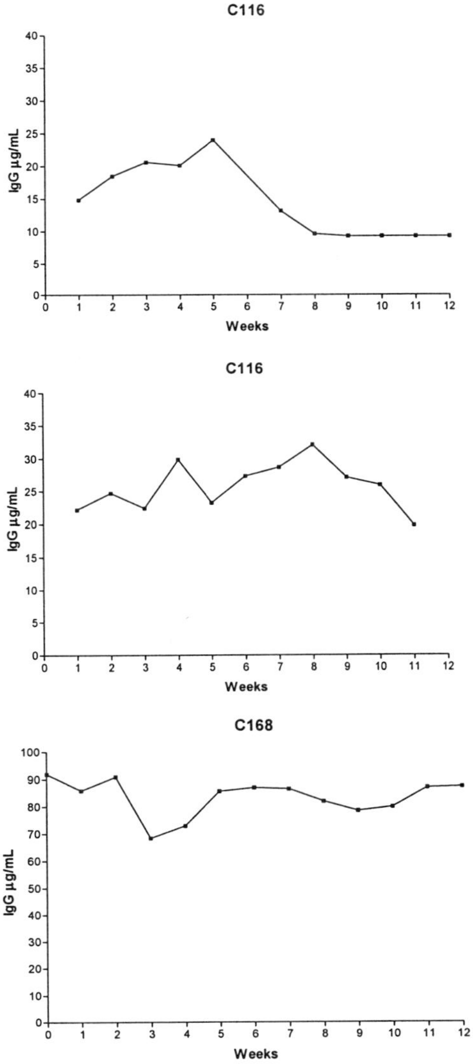


2A

2B

2C

**Figure 2.** **Three month stability studies grown with top panel: "first generation" cell line C116, middle panel: "second generation" C116 and bottom panel: cell line C168.**

Figure 4 shows the relative concentration of antibody for the same five bioreactors as a function of time. The antibody concentration increased at similar rates in all reactors and was maintained at approximately the same level during the production phase up to and including day 35. These data demonstrate that the continuous perfusion cell culture process was adequately controlled.

**Table 1 Cumulative generations from thaw of MWCB to end of bioreactor run**

| Culture | MDT (hr) | Days | Generations |
|---|---|---|---|
| Preculture | 20-30 | 12 - 15 | 11.8 - 15.1 |
| Back Up Preculture | 20-30 | 0 - 10 | 0 - 11.8 |
| 50 L Expansion | 50-80 | 3 - 5 | 3.0 - 3.5 |
| 500 L Expansion[1] | 50-80 | 9 - 12 | 5.4 - 8.1 |
| High Cell Density | 180-240 | 16 - 49 | 2.0 - 3.8 |

[1]Time in 500 L bioreactor until HCD control

A second chimeric IgG1 producing cell line, C168 was developed and validated. This process was maintained in culture for 83 days (Figure 5), and in a bioreactor for over 70 days. Cellular viability was consistently high during preculture and in the bioreactor during which time 8 kg of product was produced. Stability studies clearly demonstrated long term productivity of this cell line grown in both small and large scale cultures maintained in serum free medium (Figure 2C and Figure 6 respectively). Productivity was constant for at least 70 days. These data combined with process economics were required to establish production limits. During the entire 83 day cell culture production process (11 days preculture and 73 days post 50 L inoculation), the total cumulative generations were 23.6 (11.1 and 12.5 for preculture and production cell culture respectively).

**Stability During Production**. Table 2 shows the results of the cell bank testing for the C116 cell line. Gross genomic structure, cell line authenticity, and microbial and viral contaminants were tested on pre (MCB/MWCB) and post production cell banks (EPC/LECB) in accordance with the United States Food and Drug Administration (FDA) *Points To Consider In The Characterization Of Cell Lines Used to Produce Biologicals* (1993), the FDA *Points To Consider In The Manufacture And Testing Of Monoclonal Antibody Products for Human Use,* (1997), the Committee for Proprietary Medicinal Products (CPMP) *Notes For Guidance On Production And Quality Control Of Monoclonal Antibodies* (1994), The International Conference on Harmonization (ICH) *Guideline On Quality Of Biotechnological Products: Viral Safety Evaluation* (1997) and the ICH *Guideline On Quality Biotechnological Products: Cell Substrates* (1997). In addition an ELISA was performed on selected cell banks to establish how many cells were positive for IgG1.

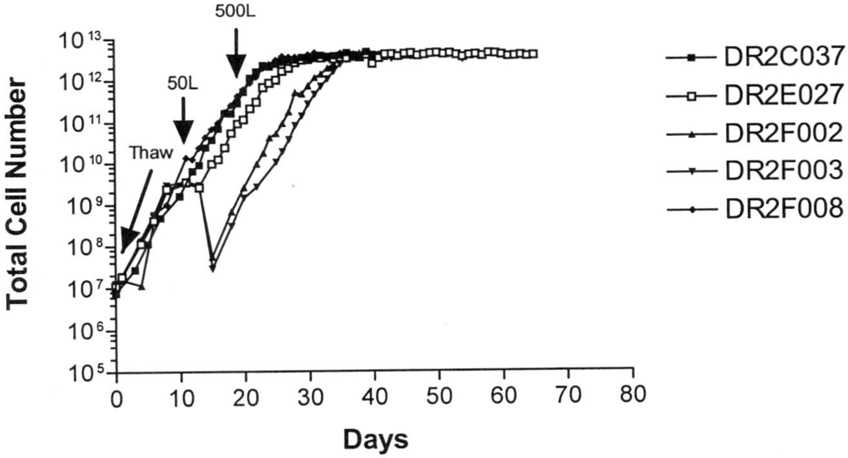


**Figure 3.** **Growth of the cell line C116, 500 L production cell cultures from thaw of the MWCB, preculture and until HCD is reached in the production bioreactor.**

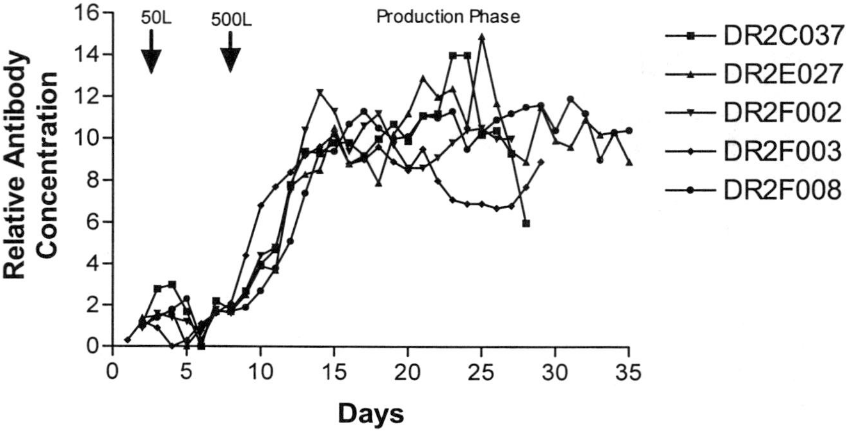


**Figure 4. Antibody titres of cell line C116 during 500 L production scale bioreactors.**

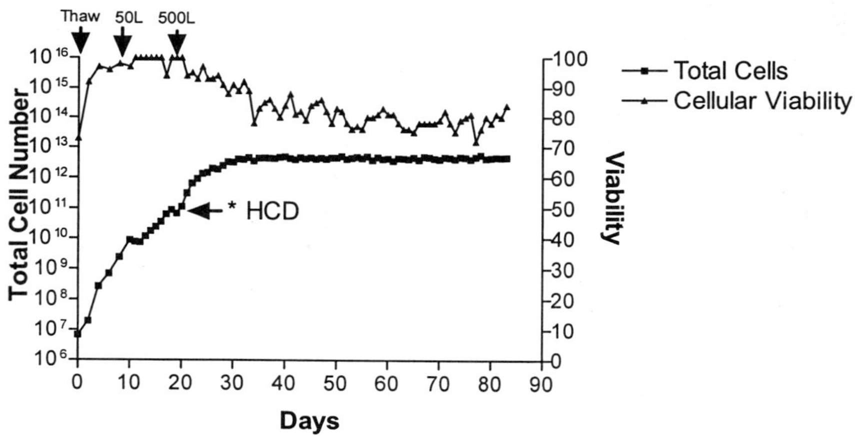


**Figure 5. Growth profile of C168 cells from thawing a MWCB vial to a predetermined maximum cell density in a 500L production scale bioreactor. The number of culture days is compared to the log of total cells.**

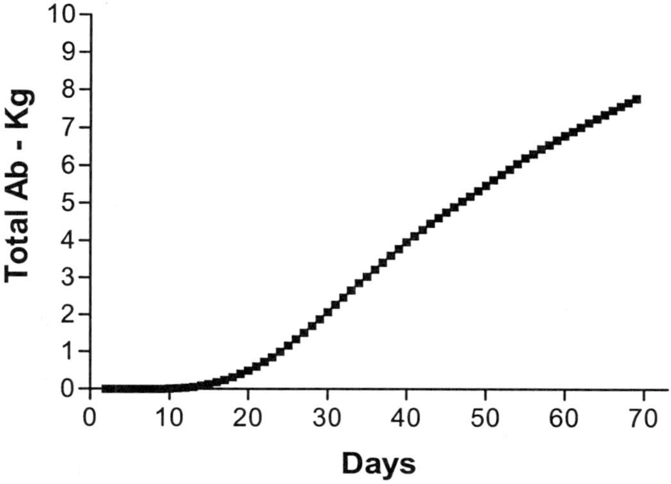


**Figure 6.** **Cumulative antibody (kg) of Cell line C168, 500 L production culture producing 8 kg antibody.**

---

**Table 2 Cell Bank Testing**

- No adventitious agents
- Endogenous Retrovirus Characterization
- Mycoplasma Negative
- Sterile

---

**Table 3 C116 Cell Bank Clonality**

| Sample | % Producers |
|---|---|
| MCB | 100 |
| MWCB | 97 |
| DR2E027 D34 | 98 |
| DR2E027 D48 | 88 |
| DR2F003 D35 | 100 |
| DR2F008 D35 | 98 |

Normal cellular viability was expected to be in excess of 70% during preculture and normal MDT for serum-containing C116 and C168 cultures ranged between 17-31 hours (data not shown). Table 3 shows the MDT and cellular viability for C168 when switched from serum-containing growth medium used in preculture to the serum free growth medium used in the perfusion bioreactor. The average MDT of 26.4 hours in serum free medium is very similar to the average MDT seen in serum-containing medium (21.6 hours) and well within the expected range for serum-containing cultures.

**Table 4 C168 preculture characteristics in serum containing and serum free medium**

| Cell Line Condition | Average Cellular Viability | Range | MDT (hr) | Range |
|---|---|---|---|---|
| Serum Containing | 87% | 73 - 97% | 21.6 | 16.8 - 26.4 |
| Serum Free | 97% | 96 - 98% | 26.4 | 19.2 - 32.6 |

Product was isolated throughout one C116 serum-containing continuous perfusion production run, and characterized to demonstrate stability. Key analytic data obtained with C116 are shown in Table 5. Samples were obtained at the beginning (Day 14), middle (Day 25) and end (Day 35) of the 35 day C116 production run. These data show that the molecular weight by mass spectroscopy, and the immunoreactivity by ELISA were unchanged throughout the bioreactor run. Figure 7 shows the tryptic peptide maps of the same 3 samples. All samples showed comparable distributions of peptides throughout the perfusion process.

**Table 5 Reproducibility of C116 mass spectroscopic analysis; product consistency within a single batch bioreactor run.**

| Sample | MW by Mass Spec | Immunoreactivity % |
|---|---|---|
| Day 14 | 47,620 | 97 |
| Day 25 | 47,617 | 99 |
| Day 35 | 47,613 | 99 |

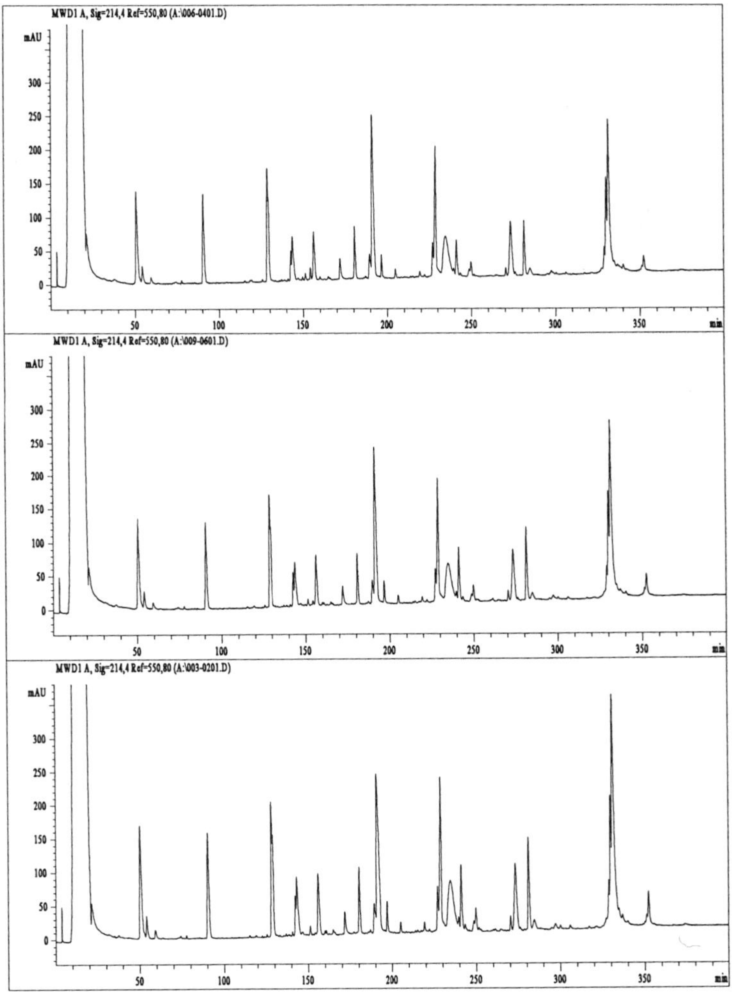


**Figure 7 Tryptic peptide maps of the Fab antibody. Antibody produced at various times during the continuous perfusion process. Panel A: Day 14, Panel B: Day 25, Panel C: Day 35.**

## Discussion

Cell lines to be used in long term continuous manufacturing processes must be homogenous, with 100% of the population producing the required antibody isotype, and the cell banks created from them must be carefully established so that each vial will give rise to an identical culture. Chuck and Palsson, (*18*) have shown that the population balance of producing and non-producing hybridomas is very sensitive to serum level, state of inoculum and medium composition, and concerns over the difficulty of creating cell lines with suitable long term stability as required for a continuous perfusion process have been voiced (*19*). We have demonstrated that it is possible to develop cell lines suitable for long term growth and high viability and to be stable for a large scale, long term continuous perfusion process. We consider a cell bank stable if specific antibody production does not decrease by more than 30% in 40 generations. Stability studies have been conducted on all MCB and MWCB cell banks for about 12 weeks (about 100 generations) to confirm the productivity (Figure 2) and growth rates (data not shown) of cell lines.

As a result of the validation program, cellular mean doubling times (MDT) and viability specifications were established in order to prevent sub optimal performance of the cell cultures especially when switching from serum containing to serum free medium. Normal viability is expected to be in excess of 70% for both preculture and production phases for all cell lines (see Figure 5 for a C168 example) and normal MDT for serum containing cultures range between 17-31 hours and 19-33 hours for serum free cultures. When C168 cells were switched from serum containing to serum free medium, no significant changes in viability or mean doubling time were observed (Table 4).

The media used in the growth phase of cell culture and the media designed to enhance antibody productivity during production phase in the bioreactors have been formulated to preclude the need for cell line adaptation or weaning especially in the absence of serum. C116 does not require selective pressure but C168 does require the presence of mycopohenolic acid in media to support high productivity levels (data not shown). For production in a continuously perfused system one may have to include a selective reagent to continually apply selective pressure and prevent genetic drift by preferential selection of a subpopulation of non-producers (*20-23*).

Validation studies were performed by monitoring accumulated generations throughout the process to show consistent cell doubling times and antibody production profiles, and to demonstrate that a consistent product was produced. During preculture, viability, cellular MDT and the number of generations accumulated were monitored and shown to be consistent (Table 1).

The stability of the cells during production is not best determined by the calculation of cell generations as the MDT increases up to 240 hours at HCD and very few generations are accumulated (Table 1). While the overall number of accumulated generations is important, a more sensitive way to monitor the progress of the reactor and, therefore, stability of cells during continuous perfusion production is by daily monitoring of cell density, viability, antibody production levels and specific antibody production. For the cell line C116, reactors have been maintained for up to 63 days of continuous perfusion. Thirty-five days from inoculation was determined to be the preferred length of the production phase which is well within the demonstrated stability of the cells during production phase. All subsequent bioreactors run in the 5 years since process validation of this cell line show consistent results.

Consistency of antibody production was also demonstrated by isolating and characterizing antibody produced throughout the production run of a single bioreactor (Table 5). In addition, EPC and LECB were established and characterized. For each bioreactor, endogenous retrovirus was characterized throughout the production process and no increases were observed when compared to the MCB and MWCB. Extensive testing was additionally performed on both the EPC and LECB to show that other viruses had not been induced nor introduced and the expression levels are consistent with the MCB and MWCB (data not shown). In addition, the cells were shown to produce stable amounts of antibody for the duration of the production process. Process validation clonality studies have shown that $\geq$ 97% of the C116 cells in the MCB, MWCB and 4 LECBs produced human IgG when made from reactors run for a maximum of 35 days (Table 3).

Process validation is a requirement of Current Good Manufacturing Practices (cGMP) regulations and is usually defined as a documented program to demonstrate that a process will consistently produce a product meeting its predetermined specifications and quality attributes. Strict control is employed throughout manufacturing according to 21 CFR parts 210 and 211; *cGMPs for Finished Pharmaceuticals* and *FDA Guideline On General Principles Of Process Validation.* In addition to the experimental studies described above, the validation program should contain a program for vendor certification to ensure a continuous supply of raw materials meeting predefined quality specifications (*24*). cGMPs require documented verification that during equipment installation qualification (IQ), there is strict adherence to manufacturers recommendations and intentions; operational qualification (OQ) ensures the equipment works as it is supposed to over operating ranges, and performance qualification (PQ) ensures that the minimum and maximum ranges are operational for all equipment associated with cell banking, preculture and the bioreactors.

## Conclusion

In summary, the selection of cell lines is of key importance and establishment of stable well characterized cell banks for use in continuous perfusion processes. Process validation generates data to establish the normal operation ranges of the process. Analysis of process intermediates is performed to confirm consistent production of chimeric antibodies. Perfusion cell culture requires that cells maintain high viability and stability for extended periods of time in a bioreactor. Sufficient in-process controls must be present to demonstrate consistent performance. Process validation studies should demonstrate: 1) that the product is unchanged throughout the production cycle, and 2) that the cells are stable throughout the production cycle.

## Acknowledgments

Thanks to S. Beckman and S. P. Wensel for their support in preparing this manuscript.

## Literature Cited

1. Avgerinos, G.C.; Drapeau, D.; Socolow, J.S.; Mao, J.; Hsiao, K.; Broeze, R.J. *Biotechnology* 1990, *8*, pp. 54
2. Himmelfarb, P.; Thayer, P.S.; Martin, H.E. *Science.* 1969, *164,* pp. 555
3 Deo,Y.M.; Mahadevan, M,D.; Fuchs, R. *Biotechnol. Prog.* 1996, *12* pp. 57.
4. Jan, D.C.H.; Emery, A.N.; Al-Rubeai, M. 11th ESACT Meeting, Brighton, UK. September 1991
5. Whitaker, S.C.; Francis, R.; Siegel, R.C. Paper 287. 213rd ACS National Meeting, BIOT Division, San Francisco, CA. April 1997
6. Griffiths, J.B. In *Animal Cell Biotechnology*, Spier, R.E. and Griffiths, J.B.Ed.; Acad. Press, London 1988, Vol 3 pp.179-220
7. Kadouri, A.; Spier, R.E. *Cytotechnology,* 1997, *24,* pp. 89
8. Arathoon W.R; Birch J.R. *Science* 1986, *232*, pp. 1390
9. Birch, J.R.; Lambert, K.; Thompson, P.W.; Kenny, A.C.; Wood, L.A. In *Large Scale Cell Culture Technology.* Lyderson, K., Ed.; 1987
10. Lebherz, W.B.; Lee, S.M.; Gustafson, M.E.; Ricketts, R.T.; Lufriu, I.F.; Marzal, A.C.; Flickinger, M.C. *In Vitro*, 1985, *16A,* pp. 21
11. Birch, J,.R.; Arathoon, W.R. In *Large Scale Mammalian Cell Culture Technology*. Lubiniecki, A.S., Ed.; Bioprocess Technology; Marcel Dekker, Inc. New York, NY, 1990, Vol. 10; pp. 251-270
12. Jo, E-C.; Park, H-J.; Park, J-M.; Kim, K-H. *Biotech/Bioeng.* 1990, *36,* pp.717
13. Knight, D.M.; Trinh, H.; Le, J.; Siegel, S.; Shealy, D.; McDonough, M.; Scallon, B.; Moore, M.A.; Vilcek, J.; Daddona, P.; *et al. Mol.Immunol.* 1993, *16,* pp. 1443
14. Knight, D.M.; Wagner, C.; Jordan, R.; MaAleer, MF.;Derita, R.; Fass, D.N.; Coller, B.S.; Weisman, H.F.; Ghrayeb, *J. Mol. Immunol.* 1995, *32,* 1271
15. Beyeler, W.; Thaler, T.; Clements, R. *Biotechnology,* 1990, *3*, pp. 24

16. Weibe, M.E.; May, L.H. In *Large Scale Mammalian Cell Culture Technology.* Lubiniecki, A.S., Ed.; Bioprocess Technology; Marcel Dekker, Inc. New York, NY. 1990, Vol. 10; pp. 147-160.
17. Birch, J.R.; Bonnerjea, J., Flatman, S.; Vranch, S. In *Monoclonal Antibodies Principles and Applications.* Birch, J.R.; Lennox, E.S., Ed.; Wiley-Liss, Inc. New York, NY. 1995, pp.231-265.
18. Chuck, A.S.; Palsson, B.O. *Biotech/Bioeng.* 1992, *39,* pp. 354.
19. Noe, W. (1997) Paper 132. 213rd ACS National Meeting, BIOT Division, San Francisco, CA. April 1997
20. Frame, K.K.; Hu, W.S. *Biotech/Bioeng.* 1990, *35*, pp. 469.
21. Osturk, S.S.; Palsson, B.O. *Hybridoma.*1990, *9*, pp. 167
22. Merrit, S.E.; Palsson, B.O. *Biotech/Bioeng.* 1993, *42*, pp. 247
23. Kessler, N.; Bertrand, S.; Aymard, M. *In vitro Cell. Dev. Biol.* 1993, *29A*, pp. 203
24. Willlig, S.H.; Stoker, J.R., In *Good Manufacturing Practices for Pharmaceuticals*, Swarbrick, J. Ed.; Drugs and the Pharmaceutical Sciences 78, Marcel Dekker, New York, 1997, pp. 73-77.

Chapter 4

# Establishment of Proven Acceptable Process Control Ranges for Production of a Monoclonal Antibody by Cultures of Recombinant CHO Cells

**Robert G. Gerber, Paul R. McAllister, Carol A. Smith, Thomas M. Smith, Dane W. Zabriskie, and Alan R. Gardner**

**Biopharmaceutical Research and Development, SmithKline Beecham Pharmaceuticals, P.O. Box 1539, King of Prussia, PA 19406**

A scaled down production culture model was used to establish proven acceptable ranges for production culture variables, including: the initial viable cell count, pH, % dissolved oxygen, temperature, and seed culture control settings. A factorial experimental design was used to estimate the effects of variation in these factors on culture performance, and to identify any resulting impact on the quality of a Mab product. Culture performance was evaluated in terms of a variety of measures that address growth, productivity, and crude purity of the secreted product. In cases where variation in factor settings caused significant effects on any of the performance measures, partially purified product (antibody) from cultures controlled at the extremes of the factor ranges were examined for effects on product quality as measured SDS-polyacrylamide gel electrophoresis, and isoelectric focusing (IEF) gels. Within the ranges examined, variation in the settings of each of the process variables caused changes in both process performance and the distribution of some IEF isoforms.

One aspect of a process validation package is to ensure that acceptable product quality is consistently attained when processes are performed according to specification. Typically, this is accomplished by demonstrating predictable process performance, and consistent product quality, when processes are controlled at setpoint, and at the extremes of specified factor ranges.

This study examines the effects of variation in initial viable cell count (VCC(*i*)), pH, % dissolved oxygen (DO), and temperature (T) settings on a process used to produce a monoclonal antibody in cultures of recombinant CHO cells grown in a stirred tank bioreactor.

 

## Materials and Methods

A humanized monoclonal antibody was produced using a recombinant Chinese Hamster Ovary (CHO) cell line. Seed cultures were scaled-up in shake flasks, and stirred-tank bioreactors (2 liter working volume), followed by stirred-tank batch mode production cultures (2 liter working volume).

**Process Variables and Variable Ranges Examined.** The production culture variable (factor) ranges that were examined in this study are listed below (Table I).

**Table I Process Factors and Factor Levels Examined**

| FACTORS | FACTOR LEVELS Low | FACTOR LEVELS High |
|---|---|---|
| Initial Viable Cell Count | 0.7 x setpoint | 1.3 x setpoint |
| Temperature | setpoint - 1 deg. C | setpoint + 1 deg. C |
| pH | setpoint - 0.2 units | setpoint + 0.2 units |
| % Dissolved Oxygen | 0.5 x setpoint | 1.6 x setpoint |
| Seed Conditions | as described below | as described below |

In addition to production reactor conditions, the effects of variation in seed culture settings were examined. In previous factorial experiments, the growth and viability of seed cultures were examined as a function of VCC(*i*), T, pH, and DO. Over the ranges examined, only pH and VCC(*i*) had an effect on seed culture growth and viability. Based on these results, three sets of conditions (Low, Setpoint and High) were used to prepare seed cultures (stirred-tank bioreactors) for this study. The seed cultures designated as "Low" were prepared with a VCC(*i*) of 70% of the setpoint value, and a pH value of setpoint minus 0.2 units. The seed cultures designated as "High" were prepared with an VCC(*i*) of 130% of the setpoint value, and a pH value of setpoint plus 0.2 units. Under all three sets of seed culture conditions, temperature and DO were controlled at the setpoint value.

**Study Design.** By using appropriate experimental designs, it is possible to estimate the magnitude of effects on process performance as a function of variation in a given controlled factor or combination of controlled factors. Typically, for processes that include multiple controlled factors, a relatively large number of experiments are required to make meaningful estimates of factor effects. In such cases, limited availability of large scale reactors often makes it impractical to perform all worthwhile process characterization studies at production scale. Alternatively, the use of small scale reactors allows a larger number of conditions to be examined and, as a result, more thorough analysis of process variation can be performed. Based on this rationale, a 2L reactor model was selected for use in this study (*1*).

At the very least, it is desirable to learn how variation in the control of individual factors affects process performance. To ensure this outcome, the experimental design must include more than one run controlled at each setting (high, low or setpoint) of a given factor. This will allow the effects of intentional variation in factor settings to be distinguished from the effects of uncontrolled sources of variation in process performance. Ideally, it is also desirable to learn to what extent each of the factors might interact to produce effects on process performance. To accomplish this goal, the design must also allow simultaneous variation in the settings of each factor that is being examined, with the combinations of factor settings selected in a way that will allow the effects of individual sets of factor combinations to be distinguishable.

By selecting a factorial experimental design, these goals can be readily accomplished. Based on the factors and factor ranges described above (five factors, each controlled at the extremes of its specified range), a full factorial design produces 32 possible combinations of high and low factor settings. By selecting a 1/2 fraction of the full factorial, fewer runs can be performed (16 combinations of high and low factor settings) without loosing the ability to distinguish the effects of variation in one or two factors at a time (*2*).

The 16 factorial runs were divided into two blocks of experiments which were run at separate times (8 runs per block). In addition, three set point runs were added to each block to aid in the assessment of block to block variation, and to allow convenient comparison of process performance in full scale and 2L cultures. This design can be most succinctly described as a $2^{5-1}$ factorial experiment, run in 2 blocks with 3 center point (setpoint) runs per block.

**Measures of Process Performance and Product Quality.** Effects on process performance were evaluated using a variety of measures. Product titer was examined because it was found to be a critical recovery process factor; based on recovery process characterization studies, a lower acceptable limit for product concentration had been established. In addition, product titer was examined because of its obvious relationship to supply and economic issues.

Product quality was examined with the use of silver and Coomassie stained SDS-polyacrylamide gel electrophoresis (SDS-PAGE), as well as isoelectric focusing (IEF) gel analysis. These analyses were performed on eluate from a protein A affinity resin that had been loaded with samples of conditioned medium.

Other general indicators of process performance were: final total cell count (TCC(*f*) - cells/ml), the integral of viable cells (IVC - viable cell days), average specific productivity (Ave. SP - picograms of product secreted per cell per day averaged over the duration of each run), crude purity (of product in culture supernatants - % of total protein represented by product) and culture age (Age(*f*) - elapsed time from inoculation to the final time point). These measures have no known relationships to product quality or downstream process performance, and were only examined as a means of gaining additional opportunities to observe variation in process behavior. Based on this rationale, no acceptance criteria were set for these measures. The experiments were concluded when the cultures reached a viability of approximately 28% (average final viability from all 22 runs).

In this study, acceptable process control ranges are established based on the anticipated effect of each variable individually, and not on the basis of the anticipated effects of factor interactions.

## Results and Discussion

**Effects on Product Titer.** Based on the 22 runs, factorial analysis was used to model product titer as a function of variation in the five production culture factors. The resulting seven term model explains greater than 80% (R square = 0.82) of the variation in product titer observed among non-set point runs. The model includes terms describing the direct effects of seed condition (SC), temperature, VCC(*i*) and DO, as well as terms that indicate interactions between temperature and the other factors. With the exception of DO, only terms with $p<0.05$ were included in the model; the DO term had a p value of 0.14 and was included in the model because the DO x T interaction was highly significant ($p=0.0125$).

Figure 1 examines the relative size of the effect that each term in the model has on product titer. The values represented by the bars in this figure are based on the absolute value of the standardized effect of each term; for this purpose the y-axis is an arbitrarily selected relative scale. This figure also indicates the cumulative effect on product titer as each term is added to the model; in this case the y-axis scale indicates the percent of the modeled effect that is explained as each term is added to the model.

This analysis indicates that seed condition, temperature and an interaction of these two factors account for approximately 60% of the effect described by the model (indicated by line *a*). This relationship is further examined in figure 2.

Figure 2 examines the effects of different combinations of seed condition and incubation temperature on product titer; each chart within this figure represents the results from a given combination of seed condition and temperature. Cultures prepared using "HIGH" or "LOW" seed conditions are represented on the right or left (respectively), and cultures incubated at "HIGH" or "LOW" temperature are represented at the top or bottom (respectively); SETPOINT cultures are represented in the center.

The product titer data from all 22 runs performed in this study are represented by the vertical bars; each bar within each chart represents the result from a specific combination of SC, VCC(*i*), DO, pH and T (as defined by the factorial design) at the indicated settings of SC and T. The average titer at each combination of SC and T is shown by the gray background level in each chart. It can be seen that setpoint conditions produced the highest average titers indicating that the process is run optimally at set point conditions. Over the ranges examined, and using the model described in figure 1, variation in seed conditions or incubation temperature alone are not predicted to result in unacceptably low titers. Unacceptably low titers were only observed when both temperature and seed conditions were below their setpoints (bottom left chart).

**Effects on Performance Measures.** Figure 3 illustrates the degree to which each of the process performance measures were affected by variations in factor settings. For each performance measure, this figure indicates the number of non-setpoint runs that exhibited a result which was significantly different from setpoint ($> \pm 2$ set point run

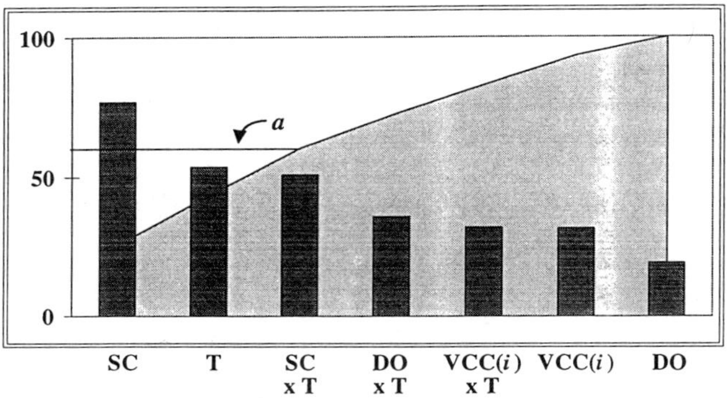


Figure 1. Relative effects of factor variation on product titer. ■ (dark grey) = the relative size of effects explained by each term, □ (light grey) = the % of total effect explained as each term is added to the model.

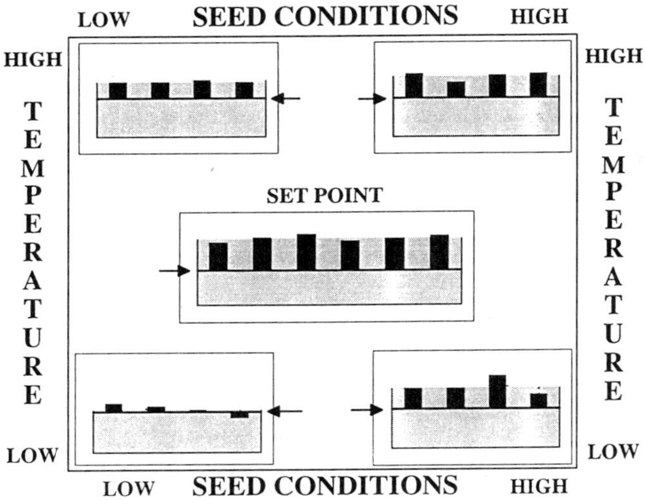


Figure 2. Effect of temperature and seed condition on product titer, □ = the average titer under the specified conditions, ■ = individual run deviation from minimum acceptable titer, ➔ indicates minimum acceptable titer.

standard deviations from the setpoint average). Fifteen of the 16 non-setpoint runs differed significantly from the setpoint average in at least one measure, and the majority of runs differed from setpoint in three or more measures.

The amount of variation in process performance that is indicated by these measures is considered acceptable, assuming that the behaviors are predictable and assuming that corresponding variation in product quality is acceptable. Because factorial experimental design was used for this study, empirical models could be fit to all of the performance measures, allowing prediction of process performance as a function of variation in the process factors (except for Age(*f*), significant models were produced for each measure- analysis not shown). This ability to predict performance is extremely useful during the transfer of processes to manufacturing: the models provide a quantitative basis for trouble shooting, optimization and for the design of full-scale process validation protocols.

**Coomassie Stained SDS-PAGE.** Figure 4 compares the purity of affinity eluate from samples of setpoint and non- setpoint runs. For this analysis, material was purified from conditioned medium by protein A affinity chromatography, and the affinity eluate was loaded onto an SDS-polyacrylamide reducing gel (equivalent protein loading per lane). After electrophoresis and staining, each lane was scanned by laser densitometer, and the percent purity (the ratio of intact heavy and light chain to total protein per lane) was determined.

Using factorial analysis, the purity values obtained from the Coomassie stained SDS-PAGE were modeled as a function of variation in the five production culture factors. The resulting seven term model (summarized in Table II) explains approximately 80% of the variation in affinity eluate purity that was observed among non-set point runs (R square = 0.79). The model has a p value of 0.0026 and includes terms describing significant direct effects of DO, and T. Terms describing the direct effects of VCC(*i*) and pH were also included in the model (although their p values were high) because these factors had significant interactions with DO, and T.

**Table II. Purity of Affinity Eluate as a Function of VCC(*i*), pH, DO and T**

| Model Satistics | |
|---|---|
| ***Term*** | ***p Value*** |
| **Intercept** | |
| **VCC (*i*)** | **0.6397** |
| **pH** | **0.2844** |
| **DO** | **0.0335** |
| **T** | **0.0595** |
| **VCC (*i*) x DO** | **0.001** |
| **pH x DO** | **0.0186** |
| **pH x T** | **0.0186** |

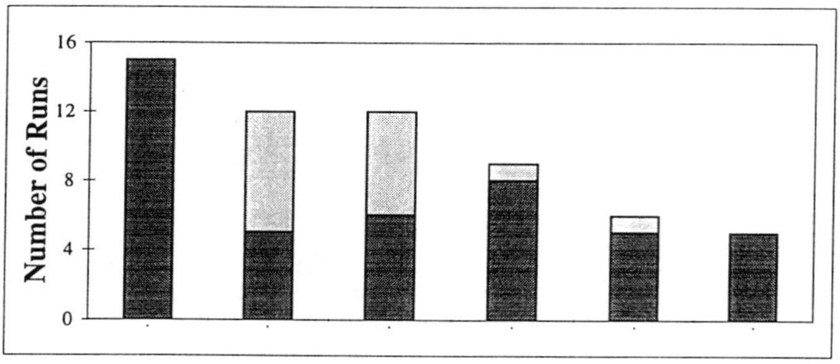


Titer TCC(*f*) IVC Ave SP Purity Age(*f*)

Performance Measures

Figure 3. Effects of factor setting variation on process performance measures. ▨ (light grey) indicates the number of runs with a response > setpoint average + 2 standard deviations, ■ (dark grey) indicates the number of runs with a response < setpoint average - 2 standard deviations.

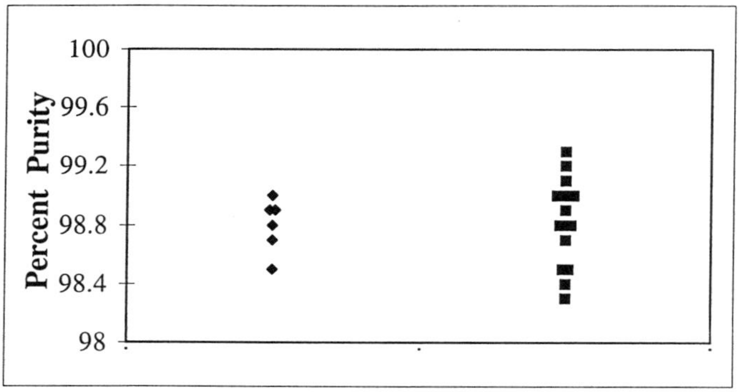


Setpoint Non-Setpoint

Figure 4. Purity of affinity eluate based on Coomassie stained SDS-PAGE. ◆= setpoint run data, ■ = non-setpoint run data.

**Silver Stained SDS-PAGE.** Silver stained reducing SDS-PAGE analysis was also performed on samples of affinity purified product from setpoint and non-setpoint cultures. Due to the inherent protein to protein variability in the efficiency of silver staining, silver stained gels were primarily examined for the appearance of additional bands in samples prepared from non-setpoint runs. Figure 5 is an example of one of the gels used in this analysis. This gel is loaded (equivalent protein loading per lane) with samples of affinity eluate from full scale setpoint cultures (FS) and set point cultures performed at 2L scale (SP). The remaining lanes are loaded with standards (D, MS and RS) and samples of affinity eluate from non-setpoint 2L cultures (X). The product bands are indicated at the right side of the gel and are labeled HC (heavy chain) and LC (light chain).

Out of the 16 non-setpoint runs (six were loaded on a separate gel), an additional band was observed under only one combination of culture conditions (arrow). Given that each high or low level of each factor is examined in 8 runs within this study, it seems unlikely that the additional band was caused by variation in any one factor (factorial analysis revealed no significant relationships between the factors examined and the presence of this band). Based on this analysis, we conclude that this finding does not represent an important effect on product quality with respect to establishment of acceptable ranges for individual factors. The only other difference that was observed between setpoint and non-setpoint samples was the loss of a band in four of the non-setpoint runs (rectangle). Because this band does not apear to be product related, its disapearance is not likely to affect the quality of the final purified material. We therefore consider this finding to be insignificant with respect to product quality.

**Iso-electric Focusing Gel Analysis.** Figure 6 compares the distribution of iso-electric focusing (IEF) isoforms in samples prepared from setpoint and non-setpoint cultures. Analysis of the IEF gels indicated the presence of 7 distinct isoforms in samples of both setpoint and non-setpoint cultures. The results are represented in terms of the percent of total protein that each isoform represents within its lane on the gel. Each rectangle in the figure contains the results from all 22 runs for a given IEF band (isoform); within each rectangle the diamonds represent the setpoint results, and the squares represent the non-setpoint results.

In the case of bands 1 and 5, there does not appear to be any difference in the range of setpoint and non-setpoint results. In the case of bands 2,3,4,6 and 7, however, it appears that variation in factor settings may cause differences in the percent of total protein represented by these isoforms. To identify any possible relationships between factor settings and the distribution of isoforms, the percent of total protein values for all seven isoforms were modeled as a function of variation in factor settings.

Table III indicates the factors and factor combinations that correlate with the distribution of a given isoform(BAND NUMBER), and indicates the significance of that correlation (X indicates $p < 0.05$, • indicates $0.05 \leq p < 0.15$). Significant models were identified for isoforms 2, 3, 4, 6 and 7, but not for isoforms 1 and 5. The extent to which each model explains the variation in the data is indicated by the R square value that is listed under each band number. Based on this analysis we

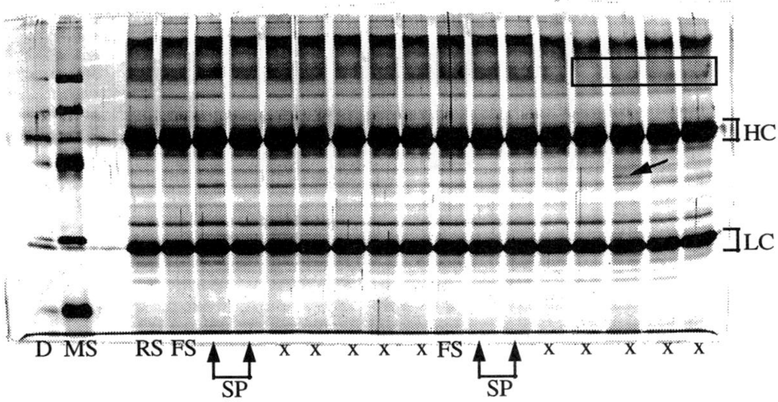


Figure 5. Silver stained SDS-PAGE. D= destain control, MS= molecular weight standard, RS=product reference standard FS= full scale set point samples, SP= small scale set point, X= small scale non-setpoint, HC= product heavy chain, LC= product light chain

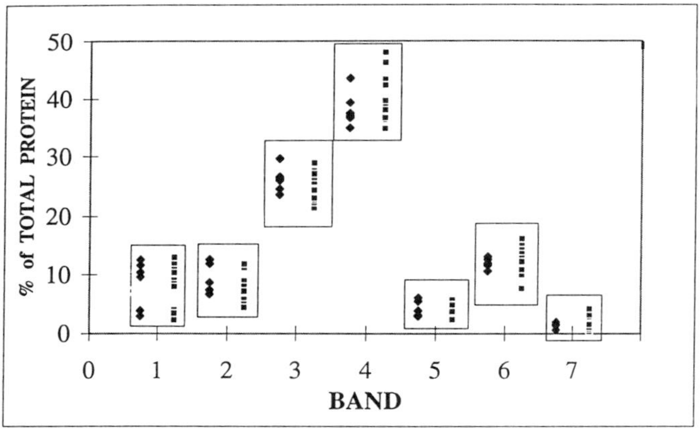


Figure 6. Effects of factor setting varition on the distribution if IEF isoforms. ◆ = setpoint run data, ■ = non-setpoint run data.

conclude that each factor examined had a direct effect on the level of at least one of the isoforms.

**Table III. Effects of Factor Setting Variation on the Distribution of IEF Isoforms**

| | Band Number | | | | |
|---|---|---|---|---|---|
| | **2** | **3** | **4** | **6** | **7** |
| *Model R Square :* | 0.34 | 0.29 | 0.48 | 0.93 | 0.92 |
| | *Terms Included in Models for each Band* | | | | |
| **VCC(*i*)** | X | X | | X | X |
| **T** | | | X | X | X |
| **pH** | ● | | | ● | X |
| **DO** | | | | X | X |
| **SC** | | | ● | X | X |
| **T x SC** | | | X | X | X |
| **T x pH** | | | | | X |
| **T x DO** | | | | ● | |
| **DO x SC** | | | | X | |
| **VCC(*i*) x DO** | | | | | X |

**X** indicates $p < 0.05$, ● indicates p between 0.05 and 0.15

In a case like this, where effects on product quality are observed, there are several actions that are worth considering. One option is to use models like those presented in this discussion, to predict factor ranges that would produce less variation in product quality. This will result in the selection of narrower factor ranges; the effectiveness of this approach will obviously depend on the ability to control the process within the narrower ranges. Another option, is to determine (if possible) the chemical identity of each isoform. Although the current level of knowledge regarding relationships between biological activity and specific chemical modifications is not likely to support a definitive assessment of risk, the collection and analysis of this type of data is likely to make risk assessment based on structural data more worthwhile in the future. As an extension of this approach, knowledge of the relationships between factor settings and product quality could be used to produce different lots of product, each containing varying amounts of the different isoforms. These materials (or preparations of pure isoforms) could then be used to examine the effect that each product variant has on the safety, potency, or stability of the final product, assuming that sufficiently sensitive assays exist.

## Conclusions

The use of a factorial experimental design aided in the development of a model which correlates variation in product titer with variation in process parameters. The most significant process parameters effecting product titer were seed condition and incubation temperature (and their product). Effects of these parameters account for more than 60% of the variation in titer that is explained by the model.

Cell culture process performance has been optimized, with parameter excursions from setpoint leading to decreased performance measures in most cases.

It was found that product titer alone is not sufficient to establish product consistency. Cultures which had acceptable titers produced product with minor variation in SDS-PAGE and IEF patterns. Of the three measures of product quality used in these experiments, Coomassie stained SDS-PAGE and IEF analysis revealed changes in product quality as a consequence of variation in individual process parameters. Both of these analyses indicated relatively small and predictable variation in the concentration of product related bands, but did not indicate the creation of any new isoforms as a result of parameter variation. Silver stained SDS-PAGE did indicate the appearance of a new band under one combination of parameter settings, however, it is unlikely that this change is due to variation in any single parameter

Based on the predictable performance of the process and relatively small variation in product quality observed, we propose that the process parameter control ranges that were examined are acceptable for manufacture of this MAb product by the cell line examined in this study. If very precise assays of biological function existed, one could investigate the relative contribution of each IEF isoform to the biological activity of the product. Often biological assays of such high precision are not available.

## Literature Cited

1. Gardner, A.; Gerber, R.; Smith, C.; Strohsacker, M.; McAllister, P.; Zabriskie, D. "Use of Scale-Down Models in Validation of Cell Processes for Production of Recombinant Biopharmaceuticals", ACS 211th National Meeting, **1996**
2. Perry D. Haaland, P. D. *Experimental Design in Biotechnology*; Marcel Dekker Inc. NY, 1989

Chapter 5

# Validation of the Recombinant Coagulation Factor IX Purification Process for the Removal of Host Cell DNA

M. W. Leonard[1], L. Sefton, R. Costigan, L. Shi, B. Hubbard, D. Bonam, B. D. Kelley, B. Foster, and T. Charlebois

Process and Product Development, Genetics Institute, 1 Burtt Road, Andover, MA 01810

A validation study was performed to determine the capacity of the BeneFIX[2] (recombinant coagulation Factor IX) purification process to remove residual cellular DNA. A multifaceted approach was taken, employing concurrent evaluation of the amount of DNA in samples obtained directly from the manufacturing process, laboratory scale challenge studies using $^{32}$P-radiolabeled DNA spiked into column load samples derived from a manufacturing run, and fractional factorial studies to determine the robustness of the two most critical DNA removal steps. The results demonstrate that the DNA challenge to the BeneFIX purification train is predictable, and that the purification process is robust, with the capacity to consistently remove DNA to very low levels in drug substance, significantly lower than current regulatory recommendations. Based on these results, BeneFIX has been approved for sale in the US without routine lot-to-lot drug substance testing for DNA.

BeneFIX (recombinant coagulation Factor IX, rFIX) is produced by a recombinant Chinese Hamster Ovary (CHO) cell line (called FIX.1F) and secreted into the culture media. In addition to rFIX, conditioned media obtained from the FIX.1F cell culture process contains host cell derived components including other proteins and DNA. CHO cells, and other continuous cell lines (CCLs) used for production of biotechnology products, can be maintained in culture essentially indefinitely by virtue of the deregulation of genes normally controlling cell growth. The presence of these oncogenes in residual host cell DNA has, in the past, been discussed as a theoretical safety risk for biological products derived from CCLs because of concerns about the possibility of cellular transformation by this potentially oncogenic DNA. However, the consensus of scientific and regulatory opinion has been moving away from this position (*1*, and references therein). In 1987, a World Health Organization (WHO)

[1]Corresponding author.
[2]BeneFIX is a registered trademark of Genetics Institute, Inc.

consultative group recommended that the safety risk was negligible or non-existent in products that contained less than 100pg/dose of cellular DNA (*2*) and, more recently, it has been recommended that DNA levels up to 10ng/dose should be acceptable for products derived from mammalian cells (*3*).

With this changing perspective, the presence of residual cellular DNA has, instead, become more an issue of process consistency and product purity. Citing the 1987 WHO report (*3*), the Center for Biologics Evaluation and Research (CBER) recommended in 1993 that lot-to-lot testing for DNA be performed with limits established to reflect the level of purity that can be achieved reasonably and consistently (*4*). However, testing for low levels of DNA in the presence of high levels of protein makes consistent and meaningful analysis in drug substance technically difficult (*5*, *6*, and references therein). Recognizing these issues, an FDA/CBER-sponsored workshop on well characterized biotechnology products concluded that process validation could provide a more appropriate and effective means of demonstrating product consistency and purity with respect to process-related impurities, including host cell DNA (*7*). At this meeting it was proposed, and generally accepted, that appropriate validation of DNA removal could obviate the need for routine DNA testing. This concept has since gained much support both within industry and with regulatory authorities (*3*, *6*, *8*, *9*).

The general difficulties associated with assaying for low levels of DNA in drug substance are further compounded for rFIX by a number of factors: the drug substance buffer matrix interferes with many DNA analytical methods (producing problems such as false signal, poor spike recovery or intra-assay variability); the DNA in the rFIX process stream is small (less than 1kb, see below), which reduces the sensitivity of most DNA assays; finally, a typical 3,600 IU rFIX dose corresponds to approximately 5mL of drug substance, which further stretches the sensitivity requirements for a per-dose assay. Each of these factors creates significant challenges for the development of a robust and sensitive, quantitative drug substance DNA release assay. Thus in order to demonstrate an rFIX drug substance DNA concentration of less than 100pg/dose, an assay would need to be able to accurately and reproducibly detect genomic DNA at less than 20pg/mL - a value that approaches or exceeds the limit of detection for most DNA analytical methods even without the additional negative impact of the small DNA size and the drug substance buffer matrix. For these reasons, rather than perform routine lot-to-lot testing for DNA, the strategy employed for BeneFIX has been to demonstrate that the purification process consistently removes DNA to acceptable levels in drug substance.

The process used to purify rFIX from conditioned media was designed to remove DNA and other impurities, and validation studies were undertaken to demonstrate the achievement of this objective. For the DNA removal validation studies, the amount and size of residual cellular DNA entering the rFIX purification process was ascertained and the capacity of the purification train to remove this material determined. There were three separate arms to the validation package: concurrent scale studies to determine the amount of DNA in samples obtained directly from the manufacturing process; laboratory scale challenge studies to determine the capacity of individual column steps to remove DNA (and identify DNA-containing process streams); and laboratory scale fractional factorial studies to determine the robustness of the two most critical DNA removal steps. For these latter studies, efforts were made to ensure that the 32P-radiolabeled DNA spike used was appropriately representative of FIX.1F cellular DNA present in the rFIX process stream. Together, the results of these studies demonstrate that the DNA challenge to the rFIX purification train is predictable and low, and that the purification process is robust with the capacity to consistently remove DNA to acceptable levels in drug substance.

## The rFIX Purification Process

rFIX-conditioned medium is separated from FIX.1F host cells by microfiltration, followed by ultrafiltration and diafiltration (UF/DF) into the load buffer for the first chromatography column. As shown in Figure 1, the rFIX purification train is comprised of four chromatographic steps followed by a Viresolve-70 virus-retaining filtration step and a second UF/DF step, to yield rFIX drug substance.

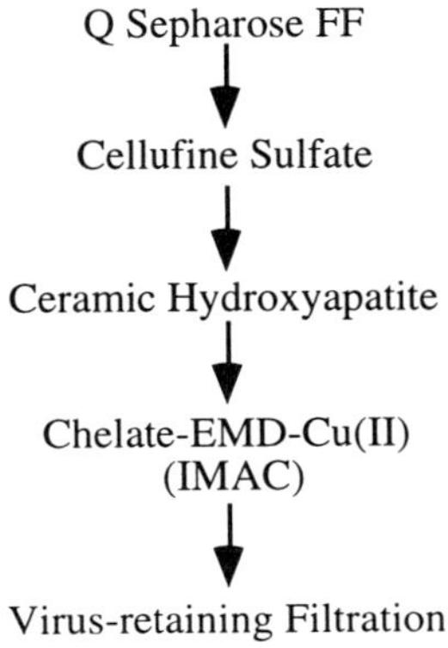


**Figure 1**. Schematic Representation of the rFIX Purification Process

## Concurrent Validation

A concurrent validation study was performed at manufacturing scale, to demonstrate that the purification process effectively and consistently removes DNA to acceptable levels. Six consecutive manufacturing batches were sampled for DNA analysis as described in an approved study protocol. All six batches met all in-process control parameters and release specifications. The results of this concurrent validation study were confirmed and extended by analysis of additional full scale manufacturing samples (in studies performed to address other specific questions).

**Evaluation of DNA in Q Sepharose Load Samples.** A quantitative comparative Southern blot method was employed to determine the size and concentration of DNA entering the purification process in Q Sepharose load samples. DNA was isolated from each sample, size-fractionated on an agarose gel, transferred to nylon membrane and hybridized to a radiolabeled FIX.1F cell total genomic DNA probe. Hybridized membranes were exposed to film (Figure 2), and the DNA signal quantitated by scanning densitometry (results are presented in Table I). The amount of DNA in each sample (Figure 2, lanes 6-11) was determined by comparison to a standard curve of similar DNA at known concentration (Figure 2, lanes 1-5), and the size of DNA was determined by comparison to a molecular weight marker. Bacteriophage-λ DNA was spiked into each sample and independently analyzed (not shown) to determine DNA recovery through the isolation procedure (Table I).

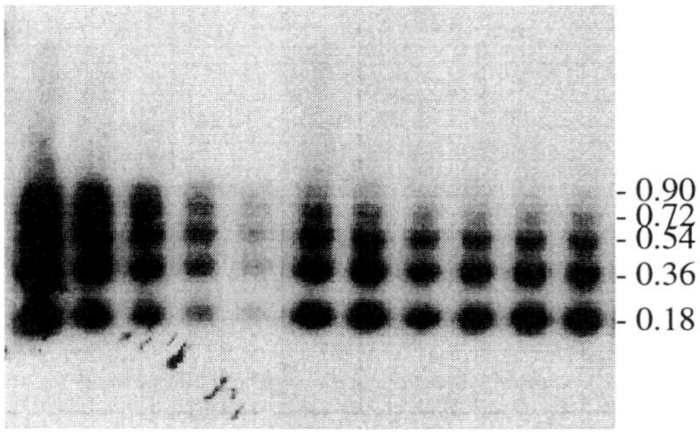


**Figure 2**. Evaluation of DNA in Q Sepharose Load Samples. DNA was isolated from the six concurrent validation Q Sepharose load samples (lanes 6 through 11) and analyzed by a quantitative comparative Southern blot assay. Following hybridization to a host cell total genomic DNA probe, the amount of DNA in each sample was determined by comparison to an oligonucleosomal DNA standard curve (lanes 1 through 5; see also Table I, below). The size of oligonucleosomal DNA fragments (in kb) is indicated.

**Table I. Analysis of Q Sepharose Load Samples**

| Sample | DNA (μg/mL) | Control Spike Recovery (%) | Total DNA Load (mg) |
|---|---|---|---|
| 5A21I013 | 1.88 | 102 | 240 |
| 5A21I014 | 1.76 | 126 | 225 |
| 5A21I015 | 0.95 | 118 | 123 |
| 5A21I016 | 1.13 | 107 | 144 |
| 5A21I017 | 1.16 | 87 | 148 |
| 5A21I018 | 1.36 | 91 | 174 |
| Average [a] | 1.37 ± 0.37 | 105 | 176 ± 47 |

[a] Values represent the mean ± one standard deviation

The results show that DNA entering the rFIX process is comprised of small oligonucleosomal fragments (less than 1kb in size), and indicate a limited and consistent extent of apoptosis in the FIX.1F cell culture process. The concentration of DNA in each of the Q Sepharose load samples was consistently low, with an average value of 1.37 (± 0.37) μg/mL, and the average DNA spike recovery was 105% (Table I). Given an average DNA concentration of 1.37μg/mL and an average volume of 128.33L, the average total amount of DNA entering the purification stream for the six concurrent validation batches was 176mg (Table I).

The concurrent validation results were confirmed and extended by quantitative Southern blot analysis of over fifty independent manufacturing scale Q Sepharose load samples. In every case the DNA size distribution was comparable to that shown in Figure 2, and the average DNA concentration was 1.88 (± 1.31) μg/mL. These results demonstrate that the DNA challenge to the rFIX purification process is predictably low.

**Evaluation of DNA in Q Sepharose Peak Pool.** The concentration of DNA in Q Sepharose peak pool samples was evaluated by DNA slot blot analysis. Samples were digested with Pronase, extracted with organic solvent, boiled to denature DNA and applied to a nylon membrane. Each sample was analyzed in triplicate with or without the addition of an appropriately-sized control spike (AluI digested FIX.1F host cell DNA), to permit determination of DNA recovery from samples. Membranes were hybridized to a radiolabeled FIX.1F total genomic DNA probe, and quantitated by PhosphorImager (Molecular Dynamics). The amount of DNA in each sample, and percent spike recovery were determined by comparison to a standard curve of AluI-digested FIX.1F DNA.

The concentration of host cell DNA in the rFIX process could not be ascertained following Q Sepharose column chromatography since it was found to be below the limit of quantitation of the DNA slot blot assay. However, in every concurrent validation Q Sepharose peak pool sample, DNA was consistently between the limit of quantitation (LOQ, 125pg/mL) and the limit of detection (LOD, 8pg/mL). The average DNA spike recovery for these samples was 99%. Given a Q Sepharose peak pool volume of 46.55L and a maximum DNA concentration of 125pg/mL, the maximum total amount of DNA present in the Q Sepharose peak pool was 5.81μg for the concurrent validation batches. This corresponds to a minimum of 4.5 log removal of DNA by the Q Sepharose column at manufacturing scale. (A DNA concentration equivalent to the LOD in Q Sepharose peak pool corresponds to approximately 5.7 log removal of DNA by the Q Sepharose column).

In every manufacturing Q Sepharose peak pool sample analyzed to date, representing over 60 Q Sepharose column runs, the concentration of DNA has always been below the LOQ and frequently below the LOD. These results clearly demonstrate the robustness of DNA removal by the Q Sepharose column under typical manufacturing conditions.

**Evaluation of DNA in Ceramic HA Peak Pool.** The rFIX product stream is concentrated during the Cellufine Sulfate and Ceramic–HA process steps and the Ceramic HA peak pool was deemed likely to be the last point in the process at which DNA might be quantitated. The concentration of DNA in Ceramic HA peak pool was evaluated by slot blot assay and found to be below the LOQ (125pg/mL), and at or below the LOD (16pg/mL in this sample matrix) in each of the six concurrent validation samples. The average DNA spike recovery for these assays was 104%. These results were confirmed and extended by analysis of fourteen additional manufacturing Ceramic HA samples: DNA was always at or below the LOD of the assay.

Given a maximum DNA concentration of 125pg/mL and a maximum volume of approximately 13.22L, the maximum amount of DNA present in the Ceramic HA peak pool for the concurrent validation batches was 1.65μg. Thus the extent of DNA clearance between the Q Sepharose load and Ceramic HA peak at concurrent scale was minimally 5.0 log in these studies. (A DNA concentration equivalent to the LOD in Ceramic HA peak pool corresponds to approximately 5.9 log removal of DNA by the rFIX process to this step). DNA could not be detected in the Chelate-EMD-Cu(II) immobilized metal affinity chromatography (IMAC) column peak pool fraction by slot blot assay.

### Laboratory Scale Validation

The results of the concurrent validation studies presented above demonstrate that the DNA challenge to the rFIX process is predictable, and that this material is consistently reduced to very low levels following operation of the Q Sepharose column. DNA is at or below the limit of detection in the Ceramic HA peak pool (the load to the final, IMAC, column step). These manufacturing scale results were confirmed and extended in a laboratory scale validation study.

**Preparation of the Radiolabeled DNA Spike**. As described above, DNA entering the rFIX purification process is predominantly less than 1kb in size. In order to generate a size-appropriate radiolabeled spike for the laboratory scale studies, FIX.1F host cell genomic DNA was digested with the restriction enzyme AluI, then 3′ end-labeled with T4 DNA polymerase and subjected to two rounds of DNA purification. This protocol generated a radiolabeled spike that was essentially free of unincorporated nucleotides, with a size distribution comparable to that of oligonucleosomal process-derived DNA present in the rFIX Q Sepharose column load (Figure 3, compare lanes 4 and 2, respectively; see also Figure 2). To ensure that the integrity of the radiolabeled spike DNA was maintained during subsequent studies, DNA was isolated from the load and DNA-containing effluent samples from each column run, and evaluated by agarose gel electrophoresis and Southern blot analysis. Representative results for one of the Q Sepharose and two of the triplicate IMAC column runs are shown in Figure 3 (lanes 5-7 and 8-10, respectively) and demonstrate that there is no detectable nuclease activity and no apparent gross size selection of DNA on these columns.

**Summary of Laboratory Scale Validation Results**. The extent of DNA removal at each chromatography step of the rFIX purification process was determined at laboratory scale using size-appropriate 32P-radiolabeled DNA spiked into a manufacturing column load retain sample. Each scaled down process step was appropriately qualified, and was treated as an independent unit operation. Triplicate runs were performed for each step according to an approved protocol which specified acceptance criteria. All column effluents were evaluated for radioactivity and summed in order to determine the total recovery of the 32P-DNA spike. The reported log removal value (LRV) for each run is calculated as the $\log_{10}$ of the dividend of the total amount of radioactivity (cpm) in the column load divided by the total amount of radioactivity recovered in the product peak pool. To determine whether DNA eluted proximal to rFIX, a prepeak and postpeak sample were collected for each column run and analyzed separately for radioactivity. The results of the laboratory scale validation, summarized in Table II, demonstrate that the rFIX purification process provides a total of approximately 9.5 log removal of DNA.

The majority of the DNA is removed on the Q Sepharose column. The average total amount of radioactivity recovered in the three runs for this step was approximately 106%, with the bulk of the radiolabeled DNA recovered in the Q Sepharose column strip fraction. The product peak pool contained an average of approximately 0.003% of the total radiolabeled cpm loaded, corresponding to an average of approximately 4.5 log removal of DNA (see also Figure 3, lanes 5-7). This is consistent with the expected chromatographic behaviour of DNA on a strong anion exchange resin.

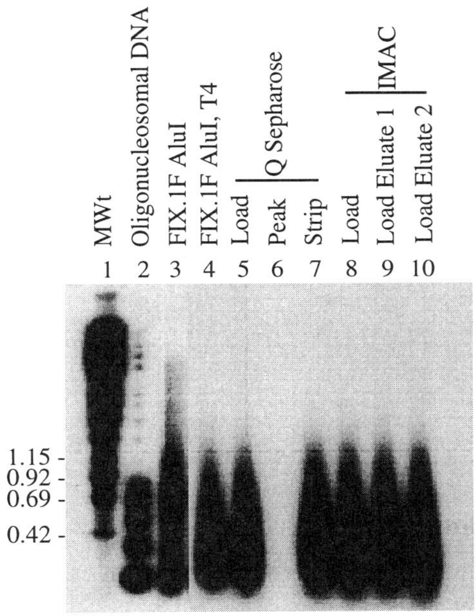


**Figure 3**. Evaluation of Laboratory Scale Radiolabeled DNA Spike. The indicated Q Sepharose and IMAC column samples were size fractionated on a 1% agarose gel, transferred to nylon membrane and detected by autoradiography prior to (lanes 4 through 10) or following hybridization to 32P-radiolabeled molecular weight marker and FIX.1F genomic DNA probes (lanes 1 through 3). The migration of DNA molecular weight marker fragments, in kb, is indicated.

The Cellufine Sulfate column provided some additional removal of DNA. The average total amount of radioactivity recovered in the three runs was approximately 95%, with the bulk of the radiolabeled DNA recovered in the flowthrough and column wash fractions. The product peak pool contained an average of approximately 4% of the total radiolabeled cpm loaded, corresponding to approximately 1.4 log removal of DNA. There was no removal of DNA observed with the Ceramic HA column. The average total amount of radioactivity recovered in the three runs for this step was approximately 92%, with essentially all of the radiolabeled DNA recovered in the rFIX product pool fraction.

**Table II. Summary of Laboratory Scale Validation Results**

| Column Step | Run #1 LRV[a] | Run #2 LRV | Run #3 LRV | Mean[b] LRV |
|---|---|---|---|---|
| Q Sepharose FF | 4.87 | 4.36 | 4.39 | 4.54 ± 0.29 |
| Cellufine Sulfate | 1.32 | 1.38 | 1.45 | 1.38 ± 0.06 |
| Ceramic HA | 0.05 | 0.08 | 0.05 | 0.06 ± 0.02 |
| IMAC | 3.52 | 3.45 | 3.70 | 3.56 ± 0.13 |
| Total | 9.76 | 9.27 | 9.59 | 9.54 |

a, Log10 Removal Value (LRV)
b, Data represent the mean ± one standard deviation

The IMAC column step provides the second greatest extent of DNA removal. The average total amount of radioactivity recovered in the three laboratory scale IMAC column runs was approximately 97%, with the bulk of the radiolabeled DNA recovered in the flowthrough fraction (see also Figure 3, lanes 8-10). The product peak pool contained an average of approximately 0.025% of the total radiolabeled cpm loaded, corresponding to an average of approximately 3.6 log removal of DNA. This is consistent with the expected chromatographic behaviour of DNA on an immobilized metal affinity capture resin.

The FIX.1F cell line has been demonstrated to be free of detectable virus and all rFIX production cell cultures are screened for adventitious virus. Moreover, the Viresolve-70 filtration step provides an additional safeguard to ensure undetectable virus in drug substance; rFIX product passes through the 70 kD nominal molecular weight cut-off membrane whereas viruses would remain in the retentate pool. A laboratory scale study was performed to evaluate the ability of the Viresolve-70 UF/DF 2 step to remove radiolabeled *Alu*I-digested FIX.1F DNA. Approximately 97 % of the radiolabeled spike DNA was found in the Viresolve-70 permeate with the product.

The results of the laboratory scale validation are in good agreement with those of the manufacturing scale studies described above. In both cases, significant DNA removal (≥ 4.5 log) was demonstrated for the Q Sepharose column. Moreover, the extent of DNA removal determined between the Q Sepharose column load and the Ceramic HA peak at manufacturing scale (5.0 - 5.9 log) is in good agreement with the sum of the removal demonstrated for the Q Sepharose and Cellufine Sulfate columns individually at laboratory scale (5.9 log). In addition, the IMAC column was shown to provide a further 3.6 log of DNA removal in laboratory scale studies (these results are summarized in Figure 4, below).

## Robustness of DNA Removal by the rFIX Process

The manufacturing and laboratory scale studies described above demonstrate that the DNA challenge to the rFIX purification train is predictably low, and that the process provides approximately 9.5 log removal of DNA. The majority of the DNA is removed at the Q Sepharose column step (greater than 4.5 log), with the bulk of the remainder (approximately 3.6 log) removed by the IMAC column. The robustness of DNA removal by these two critical steps of the rFIX purification process was evaluated using a fractional factorial study (*10*).

**Q Sepharose Column Robustness**. The Q Sepharose column is a strong anion exchanger used as a pseudo-affinity step in the rFIX purification process. Both DNA and rFIX bind to the column. Whereas rFIX is eluted by a conformation change induced at low ionic strength, DNA is essentially quantitatively recovered in the high salt column strip fraction. Thus, this column step would be expected to efficiently remove DNA from the rFIX purification stream. Consistent with this expectation, the level of DNA has consistently been below the LOQ, and frequently below the LOD, in every manufacturing Q Sepharose peak pool sample assayed to date. Moreover, the Q Sepharose peak pool DNA concentration was below the LOD even for the manufacturing batch with the highest Q Sepharose column total DNA load measured to date, representing greater than 6.0 log of DNA removal. Additionally, in laboratory scale challenge studies, the extent of radiolabeled DNA removal on the Q Sepharose column was indistinguishable in the presence or absence an 8μg/mL unlabeled DNA spike (a final, total unlabeled DNA concentration approximately 5-fold higher than the average Q Sepharose load). Together, these results clearly demonstrate that the Q Sepharose column has excess capacity for DNA removal, and that this step is very robust under typical manufacturing conditions.

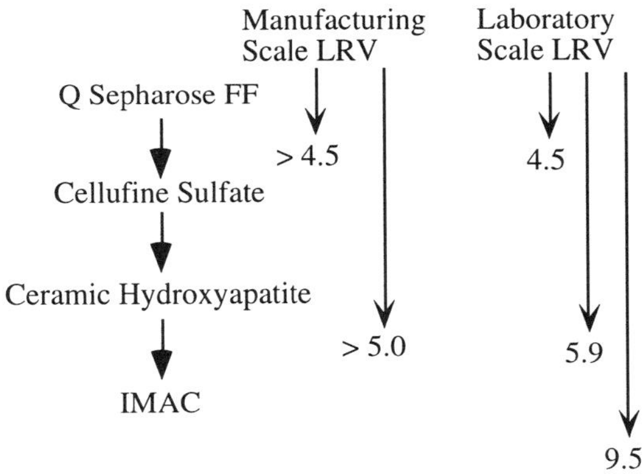


**Figure 4**. Comparison of Manufacturing and Laboratory Scale DNA Removal Validation Results. The data denote the extent of DNA removal between the Q Sepharose column load and the indicated column peak pool.

In order to further evaluate the robustness of this DNA capture step, ten critical Q Sepharose column operating parameters likely to impact on DNA removal were evaluated at minimum and maximum levels selected to reflect expected variation in manufacturing, or the limits of their control ranges (Table III), in a laboratory scale two-level, seven factor, resolution IV fractional factorial experiment. The effects of the manipulated variables were determined both singly and as confounded sets of two-way interactions.

The Q Sepharose peak pool DNA concentration was evaluated by slot blot assay for each of the 16 test and two control runs. The average DNA spike recovery for these assays was 108%. Statistical analysis of the data was hampered by the fact that all but one of the values was below the 125pg/mL LOQ of the assay. In one run a DNA concentration of 1.1ng/mL was detected. Analysis of the data indicates that such a result would only occur when several operating parameters were simultaneously at the extremes evaluated, a highly unlikely occurrence in the manufacturing suite. Moreover, even this worst case Q Sepharose peak pool DNA concentration would typically be reduced by subsequent purification steps to approximately 1.6pg/3,600 IU dose in drug substance (based on laboratory scale validation results for the steps downstream of the Q Sepharose column; see Drug Substance DNA Levels section, below). These data demonstrate that the rFIX Q Sepharose column step is very robust with respect to DNA removal.

**Table III. Q Sepharose Column Robustness Study Parameters**

| Parameter | Manufacturing Limits[a] | Robustness Study Limits |
|---|---|---|
| Load Flowrate | ≤ 2.0 cm/min | 1.6 or 2.2 cm/min |
| Load pH & Conductivity[b] | 7.8 ± 0.2<br>8.4-13.3 mS/cm | pH 7.6 @ 13.3 mS/cm<br>pH 8.0 @ 8.4 mS/cm |
| Load UF/DF 1 Volume | Not Specified | 8.3 or 16.6 mL/mL resin |
| Load rFIX Challenge | ≤ 450 IU/mL resin | 160 or 430 IU/mL resin |
| Load Viscosity | Not Specified | 2.25 or 4.50 % |
| Wash 1 Conductivity & Volume[b] | 12.6 - 17.3 mS/cm<br>≥ 5 Column Volumes | 12.6 mS/cm @ 5 CV<br>15.0 mS/cm @ 7 CV |
| Elution pH & Conductivity[b] | pH 7.9 ± 0.2<br>8.0 - 11.3 mS/cm | pH 7.7 @ 11.3 mS/cm<br>pH 8.1 @ 8.0 mS/cm |

a, Manufacturing Batch Record limit
b, Parameters were coupled for the purpose of this study

**IMAC Column Robustness**. The Chelate-EMD-Cu(II) column is an immobilized metal affinity chromatography (IMAC) rFIX capture step. rFIX product binds to the column whereas DNA does not significantly interact with the resin and is present in the flowthrough fraction. The capacity of the IMAC column to remove DNA could not be determined at concurrent scale because the concentration of DNA is at the limit of detection in the load to this step, and is undetectable in the peak pool fraction. However, in a laboratory scale validation study the IMAC column was shown to provide approximately 3.6 log removal of DNA from the rFIX purification stream under standard column operating conditions (Table II).

To determine the robustness of DNA removal at this step, seven critical IMAC column operating parameters likely to impact on DNA removal were evaluated in a laboratory scale two-level, seven factor, Resolution III Plackett-Burman fractional factorial experiment. The seven variables were evaluated in eight runs, employing radiolabeled DNA spiked into a Ceramic HA peak pool sample. With the exception of excess unlabeled DNA, high and low levels of the variables were selected to reflect expected variation in manufacturing, or limits of their control ranges (Table IV).

**Table IV. Parameters Evaluated in the IMAC Column Robustness Study**

| Parameter | Manufacturing Limits[a] | Robustness Study Limits |
|---|---|---|
| Column Equilibration Vol. | ≥ 5, ≤ 7 Column Volumes | 5 or 7 Column Volumes |
| Column Wash Volume | ≥ 5, ≤ 7 Column Volumes | 5 or 7 Column Volumes |
| Amount of DNA | Not Specified | 0 or 10 μg unlabeled DNA |
| Load pH | 7.0 - 7.4 | pH 7.0 or 7.4 |
| Load Temperature | 2 - 8 °C | 4 or 20°C |
| Load Flow Rate | 1.9 - 3.8 cm/min | 1.9 cm/min or 3.8 cm/min |
| Load rFIX Challenge | 5.37 - 8.31 $A_{280}$units/mL resin | 2.08 $A_{280}$units/mL resin<br>9.25 $A_{280}$units/mL resin |

a, Manufacturing Batch Record limit

The general performance of each of the eight runs was acceptable as measured by recovery of radiolabeled DNA (a range of 96-104% and a mean of 99%) and rFIX (a range of 105-117% and a mean of 112%). Consistent with the laboratory scale validation studies presented above, over 93% of the DNA spike was found in the flowthrough in all runs, and the column wash fraction contained the majority of the remaining DNA. Together, these two fractions accounted for 96-102% of the DNA spike, demonstrating that DNA does not significantly interact with the IMAC resin, but simply flows through the column. The extent of DNA removal was consistent, and ranged from 3.03 to 3.73 log, with an average value of 3.35 (± 0.23) log. These results demonstrate that DNA removal by the IMAC column step is robust.

Statistical analysis of the data indicates that the Equilibration #5 volume, the load pH, and the amount of DNA could impact the extent of DNA removal. Since none of the three main effects identified as significant are confounded with the two-way interaction between the remaining two variables, the simplest explanation for the model is that each main effect, although small, is significant by itself. The worst-case combination of these variables (all at their minimum values) would predict approximately 2.96 log DNA removal, while the best-case combination (all at their maximum values) would predict approximately 3.74 logs of removal.

The manufacturing and laboratory scale studies described above demonstrate that the DNA challenge to the rFIX purification train is predictably low, that the process has significant capacity for removal of DNA (approximately 9.5 log), with the majority of this DNA clearance at the Q Sepharose and IMAC steps, and that these two columns are very robust with regard to DNA removal. These results are summarized in Figure 5.

**Drug Substance DNA Levels**. DNA can not be detected in drug substance (by slot blot assay, or by the nominally more sensitive Threshold (Molecular Devices) total DNA assay). However, predicted rFIX drug substance DNA levels can be calculated by applying the log removal demonstrated for the rFIX process downstream of the Q Sepharose column at laboratory scale, to the total amount of DNA determined for Q Sepharose peak pool DNA at manufacturing scale.

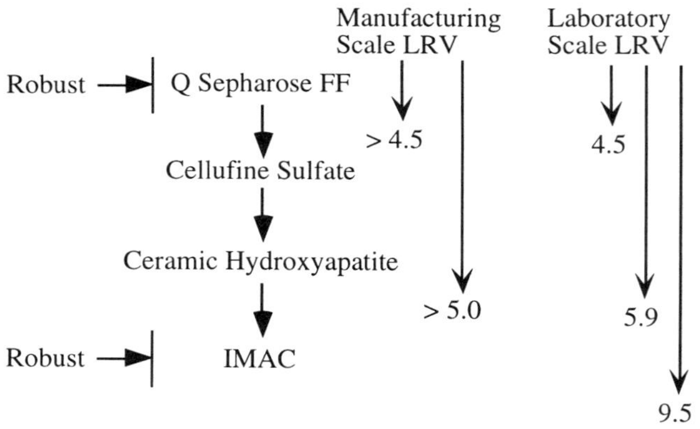


**Figure 5**. Summary of the rFIX Purification Process DNA Removal Validation Results. The data denote the extent of DNA removal between the Q Sepharose column load and the indicated column peak pool.

The concentration of DNA has consistently been less than 125pg/mL in every manufacturing Q Sepharose peak pool sample assayed to date, representing over sixty Q Sepharose column runs. Thus, 125pg/mL was taken to be a worst case Q Sepharose peak pool DNA concentration. The worst case purification process, from the perspective of theoretical levels of DNA in drug substance, was taken to be one with the maximum Q Sepharose peak volume (hence, the greatest total amount of DNA in the Q Sepharose peak pool), with the minimum extent of DNA removal on the Cellufine Sulfate and IMAC column steps (the largest total amount of DNA in drug substance), the minimum drug substance volume (the highest drug substance DNA concentration), and the lowest rFIX potency (requiring the largest drug substance volume per 3,600 IU typical dose).

For the purposes of these calculations, maximum and minimum values for the purification process parameters were defined as the mean observed for the 17 batches of a 1995 manufacturing campaign plus or minus two standard deviations, respectively. This yields values that far exceed those actually experienced in manufacturing. The worst case for DNA removal at the Cellufine Sulfate step was taken to be the mean laboratory scale validation value minus two standard deviations. For the IMAC column, the worst case DNA removal value was taken to be that identified in the fractional factorial robustness study, above. This yields Cellufine Sulfate and IMAC column LRVs of 1.26 and 2.96, respectively (values that are, again, significantly lower than the actual worst case experience in the laboratory scale validation runs).

Thus, assuming a worst case Q Sepharose peak pool DNA concentration of 125pg/mL and a worst case rFIX purification scenario described above, one can calculate an extreme worst case rFIX drug substance DNA concentration of approximately 2pg/3,600 IU. For a more typical purification scenario, with all of the parameters at their mean, this worst case Q Sepharose peak pool DNA concentration predicts an rFIX drug substance DNA concentration of 0.18pg/dose. It should be noted that DNA concentration was typically at or below the LOD (8pg/mL) in manufacturing Q Sepharose peak pool samples. At this Q Sepharose peak DNA concentration, worst case and typical case purification scenarios predict drug substance DNA concentrations of 0.13 and 0.01pg/3,600 IU dose, respectively.

## Discussion

Validation studies were undertaken to demonstrate that the process used to purify rFIX from FIX.1F conditioned media consistently and effectively removes DNA. In these studies the amount and size of residual cellular DNA entering the rFIX purification process was determined, and the capacity of the purification train to remove this material was demonstrated. The results of the concurrent scale studies demonstrate a predictable nucleic acid challenge to the rFIX purification train, both in terms of the size and amount of DNA. The Q Sepharose column consistently removes DNA to very low levels, and DNA is at or below the LOD in the Ceramic HA peak pool (the load to the final, IMAC, column step). In addition, there is good agreement between the extent of DNA clearance demonstrated in the manufacturing and laboratory scale studies. The results of the laboratory scale validation demonstrate that the rFIX purification process can provide a total of approximately 9.5 log removal of DNA. The majority of this clearance is provided by the Q Sepharose column (DNA is recovered in the column strip fraction), with the bulk of the remaining DNA removed by the IMAC column (where it is present in the flowthrough fraction). Fractional factorial robustness studies, and analysis of a large number of manufacturing scale samples, demonstrates that both of these column steps are very robust with regard to DNA removal.

Together, these studies confirm the ability of the purification train to consistently remove DNA to levels that are predicted to be, at worst case, less than 2pg/3,600 IU dose, and typically less than 0.2pg/3,600 IU dose in rFIX drug substance. These values are significantly lower than current CBER or WHO recommendations (100pg and 10ng/dose, respectively). Based on the results of this validation package, it was proposed in the US Biologics License Application and the European Marketing Authorization Application for BeneFIX that routine lot-to-lot drug substance testing for DNA need not be performed. This position has been accepted by US, Canadian and European regulatory authorities.

## Literature Cited

1. Petricciani J. C. and Horaud F. *Biologicals.* **1995**, *23*: 233-238
2. Report of a WHO study group. Acceptability of Cell Substrates for Production of Biologicals. 1987. World Health Organization Technical Report Series 747
3. Report on the WHO Expert Committee on Biological Standardization. *WHO Weekly Epidemiological Record.* **1997**. *72*, 141-145.
4. Points to Consider Document on the Characterization of Cell Lines Used to Produce Biologicals. Center for Biologics Evaluation and Research, Food and Drug Administration, Bethesda, MD, USA. 1993
5. Robertson J.S. and Heath A.B. *Biologicals* **1995**, *23*: 199-205
6. Riggin A, Davis G.C. and Copmann T.L. *Biopharm* **1996**, *9*: 36-41
7. Characterization of Biotechnology Pharmaceutical Products workshop, Washington D.C., USA. 1995

8. Points to Consider in the Manufacture and Testing of Monoclonal Antibody Products for Human Use. Center for Biologics Evaluation and Research, Food and Drug Administration, Bethesda, MD, USA. 1997
9. CPMP Positions Statement on DNA and Host Cell Proteins Impurities, Routine Testing Versus Validation Studies. The European Agency for the Evaluation of Medicinal Products. Human Medicines Evaluation Unit. London, June 1997.
10. Haaland P.D. *Experimental Design in Biotechnology*. Marcel Dekker, New York, 1989.

## Chapter 6

# Worst-Case Approach To Validation of Operating Ranges

**Alan R. Gardner, Thomas M. Smith, Robert G. Gerber, and Dane W. Zabriskie**

**Biopharmaceutical R&D, SmithKline Beecham Pharmaceuticals, P.O. Box 1539, King of Prussia, PA 19406**

A worst-case challenge is defined as executing a process under a set of conditions that leads toward process or product failure yet does not result in failure. The worst-case approach is attractive in that it allows examination of all of the critical process variables together, thus ensuring that additive effects and interactions are tested for. However running the process in this fashion poses a high risk of product or process failure. A more cautious approach is to first examine the effects of individual variables in separate experiments, e.g., in a factorial design, and then use that information to pose a worst-case which is tested experimentally. This approach was tested with a process for production and purification of a monoclonal antibody from cell culture. Process performance (antibody production and purification yield) was found to be reduced under worst-case conditions but was still within acceptable limits. Product quality characteristics were indistinguishable from product manufactured under standard conditions with the exception of the distribution of isoforms on isoelectric focusing gels. While the worst-case run demonstrated that the process can be successfully run with multiple variables at worst-case settings, it is doubtful that the same outcome would have been attained in the absence of prior knowledge of suitable ranges of critical process variables.

Operating range validation is an important component of process validation that identifies critical variables in the manufacturing process and characterizes their effect. Typically, the goal is to validate upper and lower limits on critical operating variables such that operation of the process within these limits has no adverse effect on product purity, potency, safety, or stability. The diagram in Figure 1 illustrates the concept of a process variable, for example pH in a fermentor, and its operating ranges. Normally, a target setpoint is specified for the variable, denoted as "T". In practice, however, exact control of the variable at the setpoint is never achieved; rather, the variable is usually maintained within the range known as the "normal operating range". The limits of the normal operating range, known as "alert limits" (*1*), can be established by trending process performance over the production of clinical trial or qualification lots of a product. To cover occasional excursions outside of the normal operating range, it is desirable to establish a wider range known as the "maximum operating range", within which product quality attributes are shown to be acceptable. The limits of the maximum operating range, known as the "action limits" (*1*), are established through process validation studies or operating experience. Somewhere beyond the maximum operating range is the edge of failure at which process performance or product quality becomes unacceptable, however, it is not the objective of process validation to determine the edge of failure.

While the concept of operating range validation has been applied to traditional pharmaceutical manufacturing operations such as blending and tabletting, it has only recently been explored in the context of bulk biopharmaceutical manufacturing. As shown in Figure 2, the process for manufacture of a bulk biopharmaceutical can be extremely complex, consisting of numerous unit operations, each having multiple critical input variables. Thus, the task of identifying and characterizing the effects of the critical variables can be challenging and time-consuming. The situation is further complicated when interactions between variables are considered, both within a step and between steps. Additional constraints placed on the organization performing the validation studies are that limited time and resources are available to complete the validation effort. Finally, for the validation exercise to be meaningful, it is desirable to have reasonable degree of statistical confidence in the outcome, implying that some form of replication should be built into the design of the studies.

**Worst-Case Approach.** One approach to process validation is the so-called "worst-case" approach. The worst-case is defined in the FDA's *Guideline on General Principles of Process Validation* (*2*) as "a set of upper and lower processing limits and circumstances... which pose the greatest chance of process or product failure when compared to ideal conditions. Such conditions do not necessarily induce product or process failure." The worst-case approach is attractive for validating operating ranges in a complex process containing many critical variables because potentially only a single run is required to examine all critical variables at their worst-case setting. In addition, both intra- and inter-step interactions are accounted for in such a situation because the variables are changed simultaneously. However, the worst-case approach has a number of significant limitations such as:

- Only an upper or a lower limit can be established for a given variable in a worst-case run;
- Some knowledge (either theoretical or empirical) of each variable's effect is needed to properly design a worst-case;
- Worst-case runs represent a highly unlikely processing situation and have a high probability of failure.

**Factorial Approach.** An alternative approach to validating operating ranges is the "factorial approach" (*3*) in which one or more variables in a given process step are

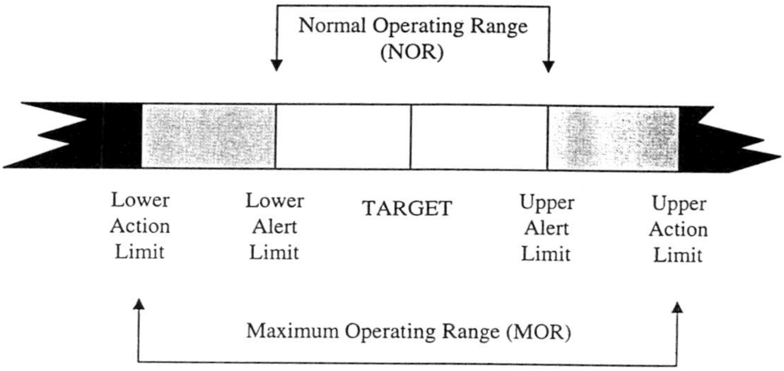


Figure 1. Operating Range for a Process Variable.

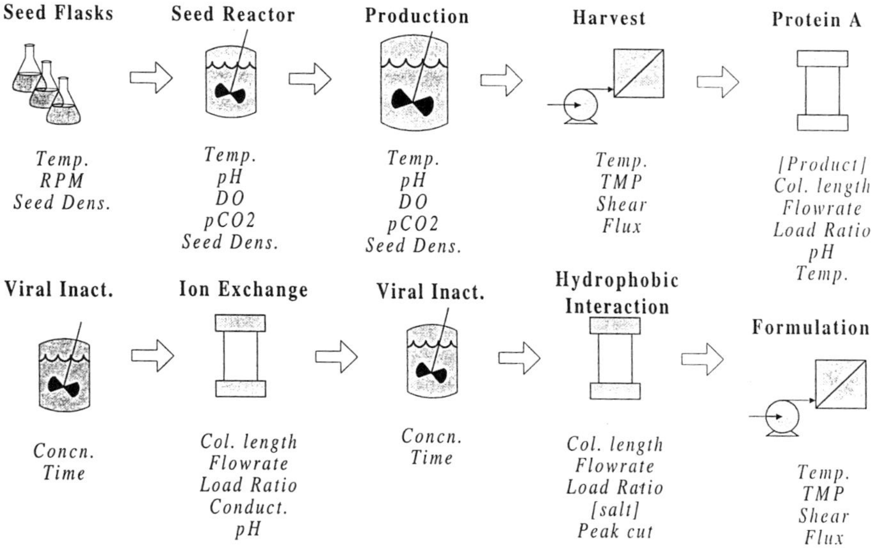


Figure 2. Typical Process Flow for Production of a Monoclonal Antibody.

manipulated in a full or partial factorial design to determine the effects on the step outcome. An advantage of the factorial approach is that, if properly designed, the effects of the individual variables and possibly some interactions between variables within a step can be ascertained.

It is possible to determine step-to-step interactions in factorial studies by including "forward linkage" variables in the design. Forward linkage variables are outputs of an upstream step that may impact the performance of the next downstream step in the process, e.g., purity of the product in the conditioned medium may impact performance or purity of the first purification step. The forward linkage variable can be handled in the same manner as any other factor in a factorial design by artificially creating feedstreams at high and low values of the linkage variable or by actually running the upstream step under different conditions to create the different feedstreams. The low and high limits of the linkage variables should represent, at a minimum, the likely ranges of the output to be encountered when the upstream step is run under its own validated conditions. If proper forward linkage variables are included for each major step in the process, then inter-step interactions will be carried throughout the entire process.

To determine whether the factorial approach with forward linkage can sufficiently predict intra- and inter-step interactions of variables, a worst-case challenge experiment was performed on a monoclonal antibody production process for which ranges of variables had been previously established via factorial experiments. In the event that comparable outcomes are achieved, it could be concluded that the factorial approach adequately considers these interactions, obviating the need for a worst-case experiment and its associated risks of failure in future validation exercises.

## Materials and Methods

A recombinant Chinese Hamster Ovary (CHO) cell line was used to produce a humanized monoclonal antibody. The cell culture process consists of seed scale-up stages conducted in shake flasks and stirred-tank bioreactors, followed by a production stage conducted in a stirred-tank bioreactor in batch mode. The purification process consists of three chromatography steps, two viral inactivation steps, and a diafiltration/ concentration step, as illustrated in Figure 2. Factorial studies were performed in scaled-down models to determine ranges for critical process variables in the major steps of the process (cell culture seed and production stages, chromatography steps, and viral inactivations) (*4,5*). Acceptance criteria for the factorial studies were based on process performance (cell growth, antibody titer, column yield), and product quality (SDS-PAGE, isoelectric focusing) appropriate to the step being examined.

The worst-case study was performed at an intermediate scale between the benchtop scale that was used for the factorial experiments and full scale. Since the benchtop scale model had been qualified to adequately represent the full scale process for the purpose of establishing process ranges (*6*), it was assumed that the intermediate scale was also a qualified model. Table I lists the variables from the production bioreactor onward that were varied from setpoint. The direction (above or below setpoint) was chosen so as to lead to poorer process performance or product quality (the latter always taking precedence). The magnitude of the variation (not shown) was determined from the maximum acceptable range determined from the factorial studies.

**Table I: Variables Examined in Worst-Case Run**

| Variable | Direction | Variable | Direction |
|---|---|---|---|
| Seed condition | – | Ion Exchange Flowrates | + |
| Seed density | – | Ion Exchange Equil. pH | + |
| Temperature | – | Ion Exchange Load pH | + |
| pH | + | Ion Exchange Load conductivity | + |
| Dissolved Oxygen | + | Ion Exchange Elution pH | + |
| Harvest Transmembrane Pressure | + | Ion Exchange Elution conductivity | + |
| Harvesting Temperature | + | Viral Inact. II Hold Time | + |
| Protein A Load Ratio | + | Hydrophobic Interaction Column Length | – |
| Protein A Flowrates | + | Hydrophobic Interaction Load Ratio | + |
| Protein A Wash Volumes. | – | Hydrophobic Interaction Flowrates | + |
| Elution pH | – | Hydrophobic Interaction gradient volume | – |
| Viral Inact. I Hold Time | + | Hydrophobic Interaction peak collection endpoint | + |
| Ion Exchange Col. Length | – | Hydrophobic Interaction [AmmSO4] | + |
| Ion Exchange Load Ratio | + | | |

+: factor run above setpoint; –: factor run below setpoint

## Results

Figure 3 shows the performance of the cell culture production bioreactor as indicated by product titer, average specific productivity, the integral of viable cell density over time, and culture age (batch time) when the cellular viability reached 60 percent. The results of four different conditions are shown: results from full-scale runs under standard conditions; small-scale runs under standard conditions (the scale being that used to conduct the factorial range experiments); small-scale runs performed under conditions close to the worst-case run; and the worst-case run. It is seen that product titer and cell growth under worst case conditions were reduced from standard conditions, while specific productivity was elevated. When compared to the small-scale runs performed near worst-case conditions, titer and culture age at 60% viability were found to agree with the prediction, but specific productivity was unexpectedly high and cell growth was somewhat low. The reason for this discrepancy is not known. Although the product titer was significantly reduced in the worst-case run, it was still within acceptable limits that allow successful recovery in the purification process based on the minimum validated loads of the full-scale columns.

Results of the purification process were compared in a similar manner for purification yield and protein purity. As seen in Figure 4, the yield over protein A was slightly lower than predicted from the scale-down run, while yields of the ion exchange and hydrophobic interaction steps were similar to the scale-down run and in fact were not significantly different from standard conditions at full scale. Figure 5 shows that purity of the process intermediates by quantitative Coomassie-stained SDS-PAGE was indistinguishable from standard conditions at full-scale in all cases. In Figure 6, the levels of protein A in process intermediates from the worst-case run were comparable to standard conditions within normal variation. These results indicate a very robust purification process.

Purified Bulk Biological Substance was characterized by a number of analytical methods normally used to release a batch. The Bulk was characterized shortly after preparation, and after one month storage at -70, 5, and 25 °C. Analysis by SDS-PAGE, Size Exclusion Chromatography, antigen binding, and cell-based potency assays all showed the Bulk to be indistinguishable from Reference Standard under all storage conditions. However, an abnormal pattern was seen on isoelectric focusing (IEF) gels in which the two most basic species were significantly higher in the worst-case samples (data not shown). To further investigate the source of this phenomenon, each process intermediate from the worst-case run was analyzed on IEF along with samples of Bulk Biological Substance from five standard runs at full scale and partially purified samples from the small-scale runs close to worst-case. The image of the IEF gel in Figure 7 shows that the shift towards the more basic species appeared in all process intermediates from the worst-case run and did not appear in any of the full-scale samples. Also, the trend was seen in small-scale samples as well. When scanned and quantitated, the shifts in isoform distribution were found to be significant in all worst-case process intermediates as shown in Figure 8. This result indicates that the phenomenon originated in the cell culture process, and that it is unlikely to be due to normal variation since it did not appear in any of the standard samples but it did appear in the small-scale samples run close to worst case. The shift towards basic pI at the small scale was also found to be statistically significant when compared to setpoint runs at that scale (*5*).

The identities of the more basic isoforms of the monoclonal antibody have not been determined. Modifications at the polypeptide level such as deamidation, cyclization of N-terminal glutamine to pyroglutamic acid, C-terminal processing, or translational read-through are known to affect charge properties that can be detected by high-performance ion exchange chromatography (HPIEC) or IEF (*7*, *8*). Changes in sialic acid content can also affect charge properties of glycoproteins. Since the amount of sialic acid on the antibody used in this study is extremely low under standard conditions (data not shown), changes in sialic acid content are unlikely to contribute to the appearance of isoforms on IEF. Therefore, the process conditions examined in the worst-case run most likely led to an increase or decrease in modifications of the polypeptide as compared to standard conditions. Before the worst-case conditions

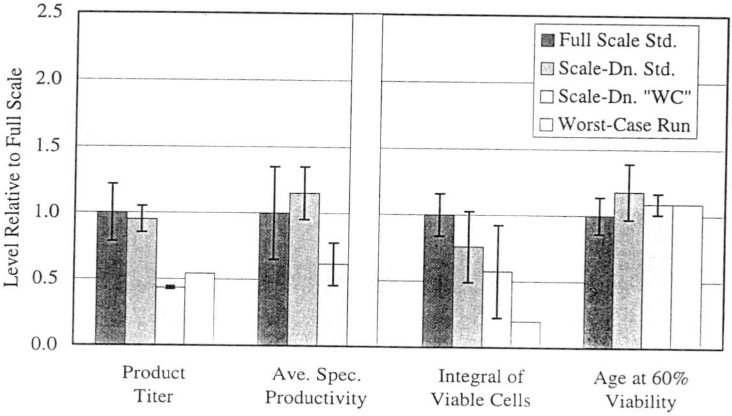


Figure 3. Performance of the Cell Culture Process. Error bars represent two standard deviations in cases where replicate experiments were run.

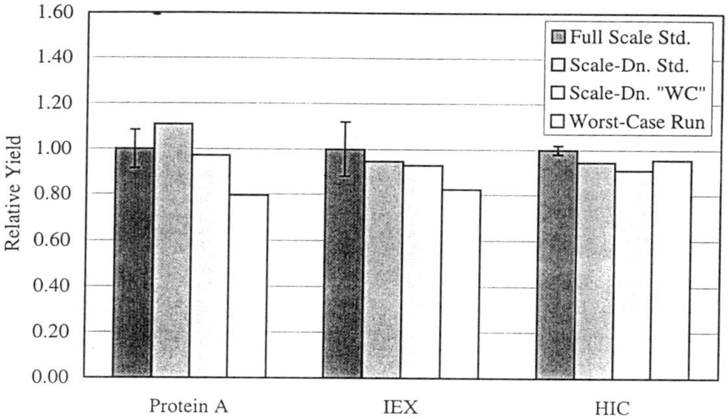


Figure 4. Yield of the Purification Process. Error bars represent two standard deviations in cases where replicate experiments were run.

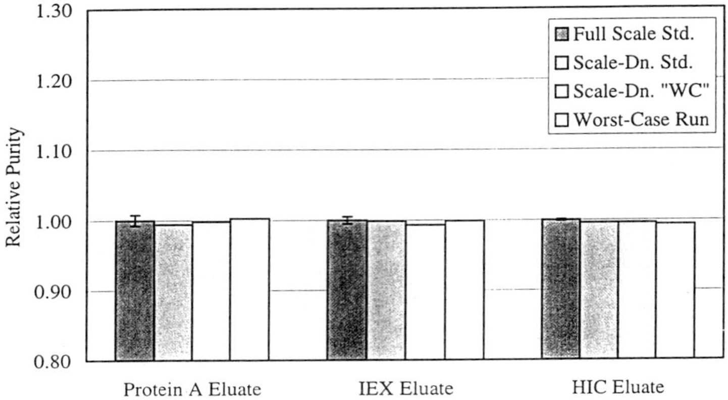


Figure 5. Purity of Purification Process Intermediates. Error bars represent two standard deviations in cases where replicate experiments were run.

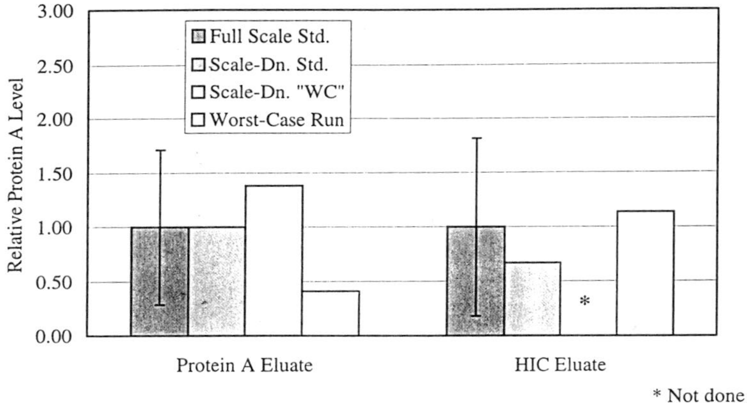


Figure 6. Protein A Levels in Purification Process Intermediates. Error bars represent two standard deviations in cases where replicate experiments were run.

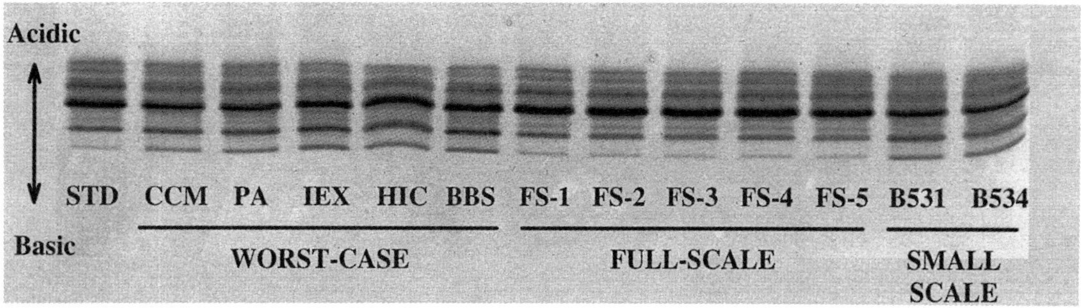


Figure 7. Isoelectric Focusing Gel from Worst-Case and Standard Conditions. STD: Reference standard; CCM: clarified conditioned medium*; PA: Protein A chromatography eluate; IEX: Ion exchange chromatography eluate; HIC: Hydrophobic interaction chromatography eluate; BBS: Bulk biological substance; FS-1 (etc): Bulk biological substance from full-scale runs at standard conditions; B531, B534: CCM from small-scale runs under worst-case conditions. *CCM samples were purified over a small-scale protein A column prior to IEF.

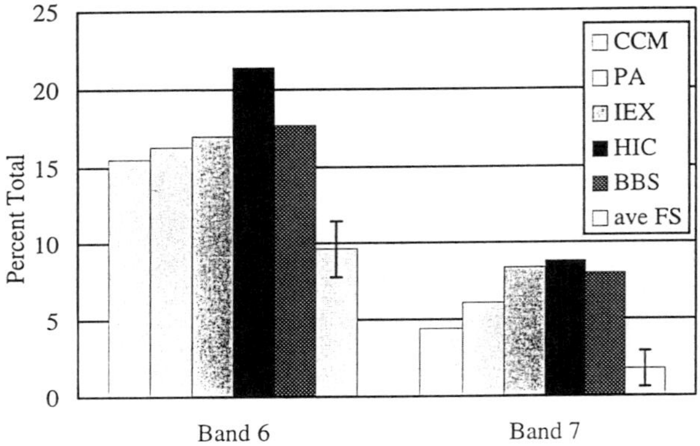


Figure 8. Quantitation of Isoelectric Focusing Gel. Error bars represent 2 standard deviations.

could be considered to be validated, these isoforms should be identified and any safety concerns should be addressed.

## Conclusions

The monoclonal antibody production process, when run with its critical variables set towards worst-case conditions, yielded acceptable performance, although product titer in cell culture and the product yield over the protein A purification step were reduced from standard conditions. Product quality was indistinguishable from Reference Standard except for the distribution of isoforms on isoelectric focusing gels. Short-term product stability was also similar to Reference Standard. It is important to note that the IEF gel did not show the appearance of new isoforms, but rather a shift in the distribution of isoforms towards the more basic charge. To consider this result acceptable, the identity of the basic isoforms should be determined and any safety concerns should be addressed.

The outcome of the worst-case run was close to that predicted from the factorial experiments that were previously conducted, suggesting that a single worst-case run could be used to accomplish what was achieved in dozens of factorial experiments. However, the worst-case run was most likely successful because the critical variables had been previously identified and appropriate ranges had been established in the factorial experiments. Had the worst-case run been conducted without any prior knowledge of appropriate ranges, the probability of failure would have been much higher. Other drawbacks of the worst-case run were that only an upper or lower limit was established for each variable examined; that effects of individual variables could not be differentiated; and statistical confidence limits were not established. Furthermore, had the worst case run failed, then there would be little guidance from the data to suggest which variable(s) were responsible for the failure and by what magnitude(s) their control range(s) would need to be adjusted to produce an acceptable result. A properly designed factorial approach can overcome all of these limitations, and inter-step interactions can be incorporated by including forward linkage variables.

## Literature Cited


1. Chapman, K. G. *Pharmaceutical Technol.* **1991**, 15, pp. 92-96.
2. US Food and Drug Administration *Guideline on General Principles of Process Validation*, 1987.
3. Kelly, B. D.; Jennings, P.; Wright, R.; Briasco, C. BioPharm **1997**, 10, pp. 36-47.
4. Smith, T.; Bodek, J.; Misczak, J.; Scott, R.; Wilson, E.; Zabriskie, D. "Establishment of Operating Ranges in a Purification Process for a Monoclonal Antibody", Chapter 7 in this book.
5. Gerber, R.; McAllister, P.; Smith, C.; Zabriskie, D.; Gardner A. "Establishment of Acceptable Process Control Ranges for Production of a Monoclonal Antibody by Cultures of Recombinant CHO Cells", Chapter 4 in this book.
6. Gardner, A; Gerber, R.; Smith, C.; Strohsacker, M.; McAllister, P.; Zabriskie, D. "Use of Scale-Down Models in Validation of Processes for Production of Recombinant Biopharmaceuticals", ACS 211th National Meeting, **1996**.
7. Jacobson, F.; Hanson, T.; Wong, P.; Mulkerrin, M.; Deveney, J.; Reilly, D.; Wong, S. *J. Chromatogr.* **1997**, 763, pp. 31-48.
8. Mulkerrin, M. "A Comprehensive Strategy for the Characterization of a Monoclonal Antibody", Waterside Monoclonal Conference, **1997**.

Chapter 7

# Establishment of Operating Ranges in a Purification Process for a Monoclonal Antibody

**Thomas M. Smith, Eileen Wilson, Robert G. Scott, John W. Misczak, John M. Bodek, and Dane W. Zabriskie**

**Biopharmaceutical R&D, SmithKline Beecham Pharmaceuticals, P.O. Box 1539, King of Prussia, PA 19406**

Validation has been receiving increasing attention in the Pharmaceutical industry. At the same time, the paradigms for validation of fermentation and purification processes are not as developed as other process stages (such as sterilization). Additionally, FDA inspection observations cite a lack of validation and establishment of operating ranges for fermentations and separations. At SmithKline Beecham, we have taken an approach that examines biopharmaceutical processes in the context of their development, scale-up and full-scale performance. We examined the combined effects of purposeful alterations in the input variables, using the concept of proven acceptable ranges and the development of small-scale models. These studies combined with full-scale data serve as the basis for establishing initial operating ranges in manufacturing.

Development of biopharmaceutical agents into marketed products has rarely progressed in a linear fashion. In an ideal world, a protein moves from discovery, through process development, scale-up to a fully validated manufacturing facility, into sequentially ordered clinical trials, and finally to FDA approval. In reality, numerous changes to the process, facilities and equipment are usually made before a biopharmaceutical enters the marketplace. At the same time, cGMPs (and indeed good scientific practice) require that processes used to produce materials administered to humans are validated. From a regulatory perspective, validation is "establishing documented evidence that provides a high degree of assurance that a specific process will consistently produce a product meeting its pre-determined specifications and quality attributes."(1) While validation has many parts, process validation in particular consists of establishing operating ranges for the critical variables in a process, and demonstrating that operating within those limits will produce a product that meets specifications.

One difficulty is determining when in the development cycle it is appropriate to invest the resources and time required for process validation. This paper presents a case study for the validation of a purification process that addresses this question by recognizing the need for process improvements during development, and serves as a basis for full-scale process validation in a manufacturing environment. By linking the development, scale-up, and validation of the process using small-scale-models, maximum operating ranges for the process were established.

## Validation Case Study: Purification of a Monoclonal Antibody

The case study describes the purification of a monoclonal antibody for therapeutic use. The purification process begins with clarified conditioned medium (CCM) from Chinese hamster ovary (CHO) cells, and ends with purified bulk biological substance (BBS). The purification consists of six processing steps (represented by the rectangles in Figure 1), including three chromatography steps (Protein A, ion exchange, and hydrophobic interaction chromatography), two viral treatments, and a formulation step using tangential flow ultrafiltration (TFUF). To reduce resin costs, the Protein A step is performed using multiple loadings per batch, with pooling of the treated eluates before the ion exchange (IEX) step. The viral treatments are included before the IEX and hydrophobic interaction chromatography (HIC) steps to address regulatory concerns associated with production in mammalian cells (2).

The focus of this case study is the establishment of operating ranges for the chromatography steps. Each of these chromatography steps is designed to perform a unique function in the process, and is operated with the appropriate setpoints for the input variables which were determined during the process development stage.

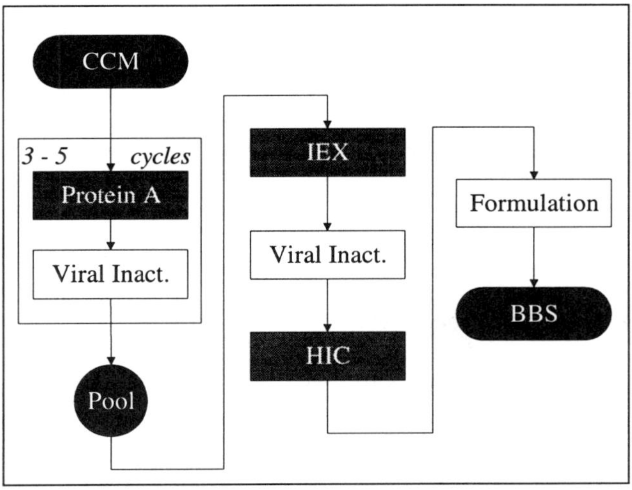


Figure 1. Process Flow Diagram for the Purification Process. CCM, clarified conditioned medium; IEX, ion exchange chromatography; HIC, hydrophobic interaction chromatography; BBS, bulk biological substance.

**Development of Small-Scale Models.** Small-scale models were designed to study purposeful alterations in the operating parameters. The design of the small-scale models followed the same general principles of scale-up: identify critical input variables, normalize variables to scale, and hold normalized variables constant across scales. To ensure the small-scale models were predictive of full-scale performance, the critical output variables were identified and measured for each chromatography step. Finally, a statistical comparison of the data was performed to compare the small and full-scale outputs.

**Identification of Input Variables.** Table I shows a partial list of critical variables for the chromatography steps. Variables such as column length, protein loading, flow rates, buffer composition, etc. are well established as influencing process performance. Other variables such as volumes of buffer applied and feed quality have direct impact on the quality of the product. The columns were scaled down by holding normalized input variables constant (e.g., column length, linear velocities, buffer compositions, relative buffer and load volumes), and using bench scale columns with diameters of 0.6 or 1.0 cm, depending on the step under evaluation. Additionally, product pooling criteria were the same between scales. Finally, feedstreams were taken from full-scale production runs to ensure the feed streams were representative and the composition constant across scales.

**Table I. Partial List of Critical Input Variables for Chromatography**

| *Variable* | *Potential Effect on...* |
|---|---|
| Column length | Yield, Purity |
| Gradient slope (normalized by bed volume) | Yield, Purity |
| Protein loading (product mass per unit bed volume) | Yield, Purity |
| Flow rate (normalized by linear velocity) | Yield, Purity |
| Temperature | Yield |
| Buffer volumes applied (normalized by bed volume) | Yield, Purity |
| Buffer composition (pH, [components]) | Yield, Purity |
| Feed quality (concentrations of HCP, protein A, product, DNA) | Purity |
| Eluate pooling criterion | Yield, Purity |

**Identification of Output Variables.** The output variables that were indicative of process performance and product quality were identified as shown in Table II. These measurements were taken to ensure that the model was predictive of the full-scale process. To identify these output variables required that the purpose of each step be clearly identified, and the data from full-scale runs was available for comparison. The yield was measured at each step, as well as the relevant quality attributes. The quality measurements were unique to each stage of the process -- not every available assay was performed at every stage in the process.

**Table II. Identification of Output Variables**

| *Step* | *Purpose* | *Performance Measurements* | *Quality Measurements* |
|---|---|---|---|
| Protein A | capture IgG | Yield[a] | Purity [b] |
| | remove host cell protein (HCP) | | [HCP] [c] |
| | remove DNA | | [DNA] [d] |
| IEX | remove HCP | Yield | Purity<br>[HCP] |
| HIC | remove protein A | Yield | [protein A] [e] |
| | remove aggregates | | [IgG aggregate][f] |
| | remove DNA | | DNA clearance[g] |

(a) by product-specific HPLC assay
(b) by SDS-PAGE densitometry
(c) by ELISA
(d) by Threshold instrument assay
(e) by ELISA
(g) Spiking and clearance experiments were required, since DNA was not detected after the Protein A step.

**Qualification of Small-Scale Models.** The small-scale models were run in triplicate and compared to a similar number of full-scale runs, as shown in Table III. As expected, the elution profiles (not shown) were similar between scales. Not suprisingly, some differences in the average output measurements were observed. However, statistical comparisons showed these differences were usually insignificant. The exceptions are noted in Table III, where $p<0.05$ in the t-test. This occurred for the yield in each step, where slight but real differences were observed between scales.

Another significant difference was in the HIC eluate, where the small-scale model eluates were on average 0.4% less pure than at full-scale. Together, these differences represent the combined effects of doing the experiments at small-scale. Having quantified these effects, the small-scale models were used to further study purposeful alterations in the setpoints of the input variables.

**Table III. Comparison of Full-scale and Modeling Data (t-test)**

| *Step* | *Output* | *Full-scale avg ± s* | *Model avg ± s* | *Difference* | *p value* |
|---|---|---|---|---|---|
| Protein A | Yield, % | 94.9 ± 4.1 (n=5) | 105 ± 0.6 (n=3) | −10.4 | 0.005 |
| | Purity, % | 98.9 ± 0.4 (n=3) | 98.3 ± 0.3 (n=3) | 0.6 | 0.10 |
| | [HCP], ppm | 7460 ± 1400 (n=3) | 9960 ± 930 (n=3) | −2490 | 0.06 |
| | [protein A], ppm | 11.1 ± 4.0 (n=5) | 11.1 ± 3.4 (n=3) | 0.00 | 1.00 |
| IEX | Yield, % | 101 ± 3.6 (n=4) | 93.3 ± 1.2 (n=3) | 7.4 | 0.02 |
| | Purity, % | 99.2 ± 0.3 (n=3) | 99.1 ± 0.2 (n=3) | 0.1 | 0.51 |
| | [HCP], ppm | 2.0 ± 0.5 (n=3) | 1.7 ± 0.1 (n=3) | 0.3 | 0.41 |
| HIC | Yield, % | 92.8 ± 0.9 (n=5) | 87.8 ± 4.1 (n=3) | 5.0 | 0.03 |
| | Purity, % | 99.5 ± 0.1 (n=3) | 99.1 ± 0.1 (n=3) | 0.4 | 0.003 |
| | [HCP], ppm | <LOD[a] (n=3) | <LOD[a] (n=3) | na | na |
| | [protein A], ppm | 0.73 ± 0.21 (n=5) | 0.43 ± 0.02 (n=3) | 0.30 | 0.06 |

(a) limit of detection
(na) not applicable

**Use of Small-Scale Models to Establish Operating Ranges.** Purposeful alterations in the critical input variables were made in combination with one another, and the output variables were measured. The experiments were designed using a "hypothesis-driven" approach: with an understanding of the factors governing chromatography, the setpoints were altered and combined to produce a "worst case" and "best case" for the output variables of yield and purity. Table IV shows the design of experiments for the Protein A step, in which the worst-case setting for the input variables was determined by their hypothesized effect on the yield. For example, an increase in the column length is expected to increase the yield by increasing the dynamic binding capacity (DBC) of the column. Therefore, a worst case setting for column length is to reduce column length.

A similar design was used for the output variable of purity, which resulted in a different worst case experiment since not every alteration in the input variables had the same effect on purity. Specifically, an increase in elution pH would selectively elute the antibody, resulting in an increase in purity. Therefore, the elution pH had a positive effect on purity, but a negative effect on yield. This called for a separate worst case experiment for purity. The best case experiments were designed by simply choosing the settings opposite to those used for the worst case. Space does not allow a tabulation of each experiment's design; the general principles described above were applied for both yield and purity outputs at each column step.

To summarize, worst and best case experiments were designed by choosing the direction of change to the input variables based on the hypothesized effect on each separate output variable. For each column step, the worst case for yield was slightly different than the worst case for purity. Thus, each step had two separate worst case experiments, and two separate best case experiments. Added to these were the three model qualification experiments, for a total of seven experiments per column step.

**Table IV. Hypothesis-Driven Design for Protein A Step**

| *Variable* | *Hypothesized Effect on Yield* | *Rationale* | *Worst case setting* | *Best case setting* |
|---|---|---|---|---|
| Column length | + | Increased DBC[a] | low | high |
| [Product] load | + | Increased DBC | low | high |
| Load ratio[b] | – | Column overload | high | low |
| Flow rate | – | Decreased DBC | high | low |
| Elution pH | – | Poor elution | high | low |
| Load pH | + | Increased DBC | low | high |
| [HCP] load | none | Affinity step | high[c] | low |

(a) DBC, dynamic binding capacity
(b) Mass product loaded per unit bed volume
(c) Either setting could be used

**Selection of Ranges for Input Variables.** After the experiments were designed, the ranges were selected based on a combination of desired operational flexibility, full-scale trending, and scientific judgment. Column lengths and flow rates were adjusted as needed; buffers and solutions were made with the desired component concentrations and pH levels.

In some experiments, the quality of the feedstreams were altered to further challenge the process. Purposeful alterations in the quality of the feedstreams were made in combination with alterations in operating conditions to allow a more rigorous assessment of process robustness. The feedstreams of the worst case Protein A and IEX experiments were spiked with host cell protein (HCP) to elevate impurity levels. Unfortunately, the levels of HCP in the normal Protein A and IEX feedstreams were not known at the time (assay under development); spiking levels were chosen based on impurity estimates by SDS-PAGE. Similarly, the feedstream of the worst-case HIC experiments was spiked with free protein A.

**Establishment of Ranges For Critical Input Variables.** For each column step, the worst and best case experiments were performed as designed, and the results were compared with the small-scale model experiments and full-scale production runs. A statistical comparison of the output variables was performed to quantify the significance of observed differences with respect to the variability observed in repeating runs. These analyses were performed by a one-way analysis of variance (ANOVA). Where a significant difference between full-scale and small-scale runs was noted, data from full-scale runs was excluded from the analysis. Figures 2 - 5 summarize the results graphically, showing p values from ANOVA only where the differences observed were significant ($p < 0.05$).

**Yield.** Figure 2 shows the yield observed in each set of experiments, along with the average yield of the full-scale runs. The Protein A step performed within a yield range of 90-105% across scales and conditions. The yield in the IEX and HIC steps ranged between 89-101% and 80-94% respectively. Within the worst and best case runs, the lowest yields were observed in the worst case yield experiments, and the highest yields were observed in the best case yield experiments. However, the differences were not always statistically significant when the data was examined by ANOVA. Figure 2 shows which experiments resulted in significantly lower yield results. For example, the Protein A worst case yield experiment was not significantly different from the full-scale average (difference = 4.1%, $p = 0.30$). However, when compared to the small-scale results, the difference was significant (difference = 15%, $p = 0.009$). Thus, the alterations designed in the worst case yield experiment significantly reduced the yield of the Protein A step when the effect of scale was considered. Taken together, these data indicate that a significant reduction in yield over Protein A may be observed in manufacturing under these conditions.

For IEX, the worst case yield experiment was significantly different from the full-scale average (difference = 12%, $p = 0.02$) However, when compared to the small-scale results, the difference was not significant (difference = 4.3%, $p = 0.26$). Thus, the alterations designed in the worst case yield experiment did not significantly reduce the yield of the Protein A step when the effect of scale was considered. Taken together, these data indicate that no significant reduction in yield over IEX is expected in manufacturing under these conditions.

For HIC, the worst case yield experiment was significantly different from the full-scale average (difference = 13%, $p = 0.003$) and the small-scale results (difference = 7.8%, $p = 0.04$). Thus, alterations designed in the worst case yield experiment significantly reduced the yield of the HIC step. A significant reduction in yield over HIC may be observed in manufacturing under these conditions.

These results indicate that yield can be compromised when worst-case operating conditions are combined. However, it is important to note that the combination of worst-case conditions which give rise to these results are unlikely to be encountered in production.

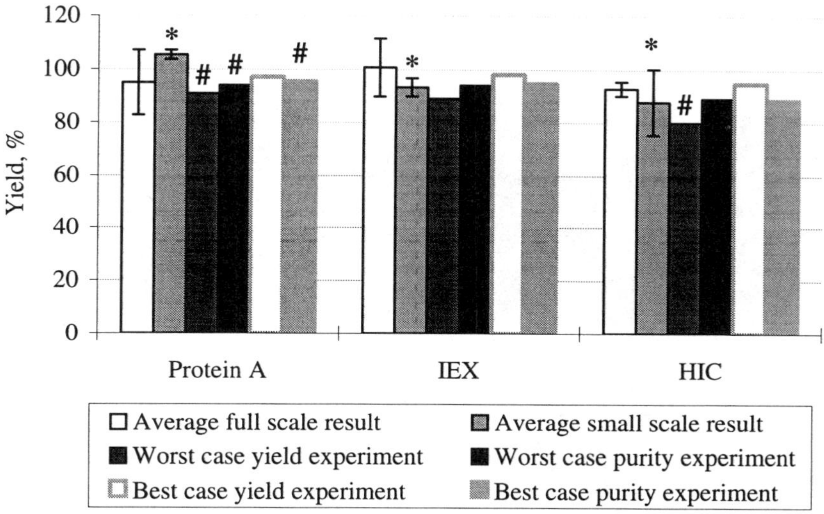


Figure 2. Yield Comparisons. Error bars represent ± 3 standard deviations; (*), significant difference between full-scale and small-scale model observed ($p < 0.05$); (#) significant difference observed in operating range experiments ($p < 0.05$).

**Purity.** Figure 3 shows the purity observed in each set of experiments, along with the average purity of the full-scale runs. The average purity of the full-scale intermediates increased as the purification progressed, as expected for a robust process. Additionally, the variation in purity decreased as the purification progressed, as expected for a robust process. This can be visualized by the relative length of the error bars in Figure 3.

The Protein A step performed within a purity range of 98.3-99.0% across scales and conditions. No significant changes in Protein A eluate purity were observed in any experimental runs. For IEX, purity ranged between 98.8-99.4%. No significant changes in IEX eluate purity were observed in any experimental runs. Thus, despite the reductions in yield observed in the Protein A and IEX experiments discussed above, the purity of the Protein A and IEX eluates was unchanged.

As previously noted, the small-scale model of the HIC step provided a slightly lower but statistically significant difference in purity (difference = 0.4%, $p = 0.003$). However, when the full-scale HIC data was excluded from the ANOVA to correct for the effect of scale, no significant reduction in the purity of the HIC eluate was evident in any experimental run. Thus, each process step was highly robust with respect to the purity of the eluates.

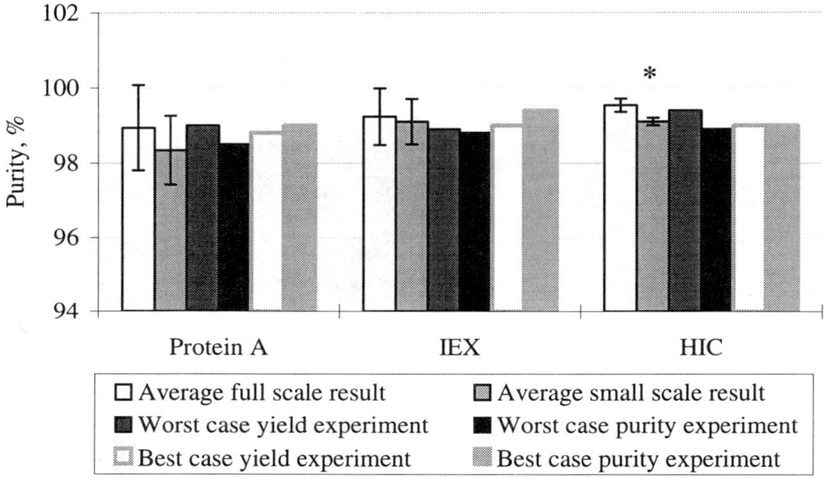


Figure 3. Purity Comparisons. Error bars represent ± 3 standard deviations; (*), significant difference between full-scale and small-scale model observed ($p < 0.05$).

**HCP Content.** Though HCP content was not included as a design component for the experiments, frozen retention samples were assayed to measure the robustness of the process by this highly sensitive impurity assay. Figure 4 shows the HCP content observed in each set of experiments, along with the average of the full-scale runs. The Protein A eluates from all the experiments ranged from 1,800-101,000 ppm of HCP. The HCP content in the IEX eluates ranged between 1.7-160 ppm. Data from the HIC experiments was excluded since the HCP content was below the limit of detection in all HIC eluate samples.

The worst case Protein A experiments involved spiking unfractionated HCP into the feedstream, in combination with operational changes. An examination of the results showed that as predicted, the experiment with the highest HCP content was the worst case purity experiment, and the experiment with the lowest HCP content was best case purity. However, the HCP content in the worst case purity experiment was not significantly higher than the full-scale runs (difference = 2670 ppm, $p=0.12$). All other Protein A experiments showed levels of HCP which were lower than the full-scale and small-scale model runs. Thus, despite elevated HCP levels and worst-case operational conditions, no change in the HCP content of Protein A eluates is expected in manufacturing.

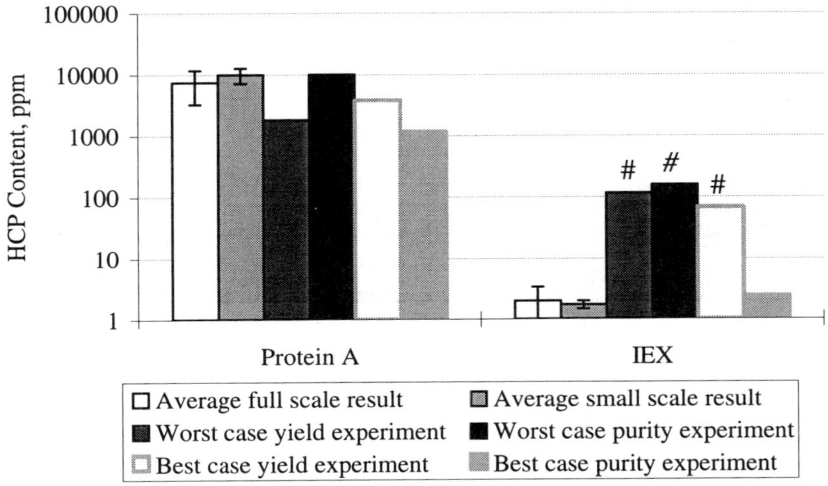


Figure 4. HCP Content Comparisons. Error bars represent ± 3 standard deviations; (#) significant difference observed in operating range experiments ($p < 0.05$). Note logarithmic scale of y-axis.

The worst case IEX experiments also involved spiking unfractionated HCP into the feedstream, in combination with operational changes. An examination of the IEX results showed that as predicted, the experiment with the highest HCP content was the worst case purity experiment (160 ppm). This level was significantly higher than that observed in full-scale runs (2 ppm) and model qualification runs (1.7 ppm). Thus, IEX experiments with elevated HCP levels in the load were observed to have elevated levels of HCP in the eluates. These results indicate that the HCP content of the IEX eluate is substantially higher when the HCP in the feedstream is elevated. However, these findings are mitigated by the following points:

1. The Protein A step was robust -- able to clear two-fold elevated levels of HCP -- and unlikely to produce a feedstream for the IEX step with elevated levels of HCP.

2. The process includes a third orthogonal chromatography step (HIC) which is likely to remove substantial levels of HCP.

3. The HCP used to spike the IEX feedstream was unfractionated and contained proteins which would otherwise have been removed. (Clearly a better experiment is to perform a "mock" purification of HCPs over Protein A prior to spiking).

Elevated levels of HCP were also observed in the IEX best case yield experiment, without HCP spiking. This experiment involved eluting the product with a high pH/high conductivity elution buffer. This buffer co-eluted HCP contaminants not normally found in the IEX eluate. Establishing appropriate control limits for the elution buffer is indicated here.

**Protein A Content.** During elution from the Protein A step, small quantities of the ligand (protein A from *Staphylococcus aureus*) leach into the product. The ligand content observed in the Protein A and HIC experiments relative to the average of the full-scale runs is shown in Figure 5. The IEX step did not remove protein A from the product, and was not evaluated here. The Protein A eluates from all the experiments ranged from 11-20 ppm protein A. There was no significant increase in the ligand leaching level in any of the Protein A experiments.

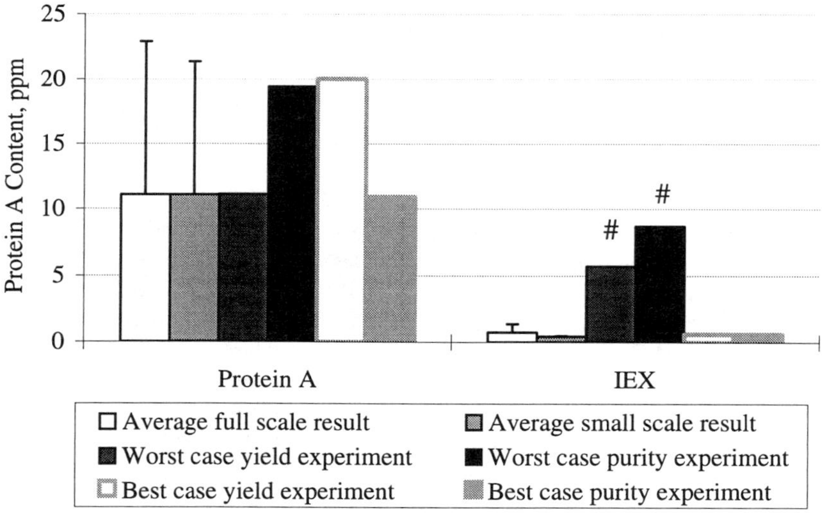


Figure 5. Protein A Content Comparisons. Error bars represent ± 3 standard deviations; (#) significant difference observed in operating range experiments ($p < 0.05$).

The Protein A content in the HIC eluates ranged between 0.4-8.7 ppm (see Figure 5). The lowest levels were observed in the small-scale model and best case experiments (0.4-0.6 ppm). Both worst case HIC experiments involved spiking protein A ligand into the feedstream at 50 ppm levels, in combination with operational changes. As predicted, these experiments resulted in the highest level of protein A ligand content. Additionally, a calculation of the clearance of protein A showed that clearance is not a constant. When protein A levels were low (5-10 ppm), clearance was high (10-20 fold). However, when protein A levels were high (50 ppm), clearance was low (6-8 fold). Finally, and most importantly, in all HIC experiments, the protein A content was acceptable when compared to product specifications.

The HIC results indicate a need to limit the amount of protein A leaching allowed in the process. This can be accomplished by raw materials testing and limiting the Protein A column lifetime to a specified number of uses.

**DNA Content and Clearance.** The DNA content of Protein A eluates and the results of DNA spike/clearance experiments in the HIC step showed no differences between full-scale, small-scale model, worst or best case runs (data not shown). Thus, the process was highly robust with respect to DNA removal.

**Conclusions.** Using small-scale models proved a useful means of establishing the effects of purposeful alterations in the critical operating parameters for the process under study. The small-scale models designed were predictive of the full-scale process, and were able to establish ranges outside the normal operating ranges for most of the input variables without compromising process performance or product quality. Where differences were significant, it revealed what changes to expect in manufacturing when input variables are altered in combination. This was accomplished with seven runs per column, studying six or more variables in each experiment. A one-fourth factorial experiment studying six variables at two levels would require at least 16 runs per column (or 64 runs for a full factorial). The hypothesis-driven approach required fewer runs, making it an attractive, effective and efficient approach. The key to the hypothesis-driven approach was the ability to form reasonable hypotheses having prior knowledge of chromatographic principles and the relationships between variables and outcomes. Without this prior knowledge, a factorial designed experiment for screening critical parameters and their effects is advised.

The operating ranges of input variables were established for the process, and serve as a basis for initial ranges in manufacturing. In general, deleterious changes in process performance and product quality were not observed in the worst and best case runs. However, there are three specific exceptions to this observation that warrant further discussion. Firstly, the yield of the Protein A and HIC steps was reduced under worst-case conditions. This yield reduction was the only observed effect; there was no measurable change in product quality. Specifically, the Protein A and HIC eluates showed acceptable purity and levels of HCP, protein A, and DNA. Such a reduction in yield may be acceptable under these conditions, provided process economics remain uncompromised.

Secondly, in the case of the IEX step, the increases observed in eluate HCP content indicates that this step can be influenced by the HCP in the feed stream and by the pH and salt content of the IEX elution buffer. A closer examination of the data revealed that elevated HCP levels in the feedstream were unlikely to be observed in production, since the Protein A step was highly robust in removing HCP. Thus, the critical observation from the IEX experiments was that the elution buffer must be tightly controlled with respect to its pH and conductivity.

Thirdly, in the case of the HIC step, an increased level of protein A in the feed stream in combination with other worst case settings resulted in higher (but acceptable) protein A levels in the HIC eluate. Thus, close control and monitoring of protein A leaching is indicated in this process. This can be accomplished by establishing limits for protein A leaching in the raw material and by monitoring protein A leaching in the Protein A or IEX process intermediate.

In summary, these findings are valuable information for defining operational limits in manufacturing operations. Ultimately, the data collected in manufacturing will complete the validation of operating ranges. Finally, the results from these experiments were used to perform a worst case run of the process from beginning to end as described in an accompanying paper entitled "Worst-Case Approach to Validation of Operating Ranges" (Gardner, et al.).

## Literature Cited

1. U.S. Food and Drug Administration; *Guideline on General Principles of Process Validation* 1987.
2. CPMP/ICH/295/95 Consensus Guideline; *Quality of Biotechnology Products: Viral Safety Evaluation of Biotechnology Products Derived from Cell Lines of Human or Animal Origin*; Geneva, Switzerland, 1997.

Chapter 8

# Robustness Testing of a Chromatographic Purification Step Used in Recombinant Factor IX Manufacture

**Brian D. Kelley, Lei Shi[1], Duane Bonam, and Brian Hubbard**

**Genetics Institute, One Burtt Road, Andover, MA 01810**

A robustness study was conducted to ensure that a chromatographic step used in the manufacture of recombinant Factor IX (rFIX) had appropriate control limits. Ten variables were tested at upper and lower limits employing a fractional factorial design requiring sixteen runs. The effects of the variables and their interactions on rFIX yield, purity, and composition were assessed by graphical and statistical methods. rFIX yield was affected by coupled elution conditions (pH, conductivity), and a two-way interaction between these variables and coupled wash conditions (pH, volume). rFIX purity was affected by load conditions (pH, conductivity), and elution conditions (pH, conductivity). A subsequent study found the elution conductivity to be the dominant variable; a revised conductivity limit was tested and implemented. rFIX composition was assessed by characterization assays, and was unaffected by the variables tested. This study provides assurance that the step will not fail when operated within the defined process ranges.

The FDA states that each step of a cGMP manufacturing process must be controlled to maximize the probability that the finished product meets all quality and design specifications and that test conditions should encompass upper and lower processing limits that pose the greatest chance of process failure (*1*). The concept of robustness is not described in the FDA's process validation documents, but is instead defined in the FDA's publication of the ICH Guidelines on Validation of Analytical Procedures: Definition and Terminology (*2*), which defines robustness as the capacity (of the process) to remain unaffected by small, but deliberate, variations in (process) parameters and provides an indication of (process) reliability during normal operation. A robust process step is one which is

[1]Current address: Amgen, 1840 Dehavilland Drive, Thousand Oaks, CA 91320.

capable of performing adequately within its control limits, consistently providing material of defined purity, quality and yield. Demonstration of the robustness of a manufacturing step prior to the start of commercial manufacturing reduces the risk of process failures resulting from inappropriate control ranges for important variables.

Establishing the robustness of a complex unit operation requires an evaluation of many process variables. It is not unusual to have as many as ten variables that require definition and control for successful operation. In addition, the potential for variables to interact with each other adds another layer of complexity, and requires the use of a test methodology capable of determining the effects of a large number of variables and their interactions. For this reason, fractional factorial designs are ideally suited for robustness studies (*3-7*). Variables are tested at two levels, representing the high and low limits of the range being tested. The interactions between variables are measured, as a consequence of the examination of all variables simultaneously. Finally, many parameters may be tested in a small number of experiments. By testing all variables at the extremes of operation, and showing these limits to be acceptable, there is a reasonable assurance of consistent operation over the full range of each variable.

Fractional factorial designs have been used for many aspects of process development in biotechnology, including media development for bacterial fermentation (*8-11*), enzymatic catalysis (*12*), analytical chromatography (*13-14*), and chromatographic resin characterization (*15*), but have not been extensively applied to purification process development. Some studies employing this type of design have been reported for cell culture optimization (*16-17*). This paper describes an approach for testing processing limits employing fractional factorial designs. The principles for the general case will be outlined, and a case study for chromatographic purification of a recombinant protein will illustrate variable selection, range setting, and statistical analysis.

## General Principles of Robustness Study Design

The philosophy and strategy of robustness study design are described in detail in an earlier case study (*18*). A brief summary is provided here.

**Identification of the Process Goals.** The goal of the process step should be clearly identified to select appropriate measurements and endpoints. This may be best defined by identifying what constitutes minimal acceptable performance. Goals should include a minimally acceptable product yield, and aspects of product quality, such as purity and consistency.

**Definition of Outputs.** Using the stated process goals, the analytical tests which measure the process performance against the stated goals should be identified. If the tests do not exist for in-process purity, one must either develop tests for the intermediate product pool, or continue the purification process to the point where an appropriate analytical test can be performed.

**Choice of Variables.** All sources of process variation should be considered as variables to be tested. The importance of any variable is integrally tied to the range over which it is tested. Extremely narrow variations decrease a variable's effects on a process, while wide variations are more likely to cause process failure. Borderline variables may be included if the fractional factorial design allows additional variables to be tested without an undo increase in the study size, or a decrease in resolution. If the effects of two variables are known to cause variation in a process output based on the same physical principle, then coupling high and low values of the two variables to give the two possible extremes represents the most severe test of robustness, while reducing the number of factors examined in the fractional factorial design, or allowing an additional variable to be tested. If the pair is important, then additional experimentation is necessary to uncouple the two variables.

**Selection of Variable Ranges.** Fractional factorial designs require high and low limits to be tested. These should represent the approximate process control limits for controlled variables, or expected limits for uncontrolled variables. If full-scale data from clinical manufacturing is available, the distribution of these populations may be used to set limits by employing a range encompassing the variables' average plus or minus two or three standard deviations.

**Experimental Design.** Various designs are possible, depending upon the number of variables to be tested, and the number of individual experiments which can be performed (often 8 - 16 runs). Each study has a characteristic resolution, which describes the ability of the study to provide information on the individual variables' effects which are confounded with two-way interactions (Resolution III), or having only two-way interactions confounded (Resolution IV), or unequivocal determination of both the variables' effects and their two-way interactions (Resolution V). This allows the experimenter to choose an appropriate design based on available resources, prior development information, and the necessary level of detail required from the study (*3, 5*).

Centerpoints should be added to the design. Centerpoints test for curvature effects, which arise from variables having a non-linear effect on a process output. Such centerpoints, when replicated, may also serve to provide information about the reproducibility of the process and associated assays. This information can prove useful in analysis of the data.

**Analysis of Experimental Results.** The analysis of fractional factorial designs can be performed by several different methods (*3, 5-7*). Frequency histograms indicate the distribution of all outputs. Pareto plots provide a rank ordering of the variables' effects, and estimate the relative strength of each variable or two-way interaction. ANOVA determines the statistical significance of the variables' effects, and should be used to define the most appropriate model predicting the output. The choice of appropriate models may employ various methods of model building using stepwise addition or elimination methods (*19*).

## Materials and Methods

**Chromatography.** All chromatographic runs were performed on an automated LC system using a qualified, validated scaled-down version of the rFIX manufacturing step (*20-22*). The chromatographic step consists of an eight column volume (CV) load, two low-conductivity washes of five and three CVs, and a step elution using a calcium-containing buffer.

**Variable manipulation.** The robustness study required that eight different types of column loads be prepared with various combinations of high and low levels of rFIX concentration, polyvinyl alcohol (PVA) concentration, and load conductivity and pH. The standard load material was diluted 1:1 with the diafiltration buffer with or without added PVA. These high and low PVA intermediates were then spiked with purified rFIX as needed to give the required rFIX concentrations. Finally, these four different intermediate pools were titrated and NaCl stock solution added to manipulate the pH and conductivity, yielding eight different load combinations.

**Analytical methods.** rFIX concentration was measured by a clotting assay employing the Coag-a-mate assay according to internal Standard Operating Procedures. The purity of the rFIX pool was measured by a host cell protein ELISA test. Size exclusion HPLC was used to determine aggregate levels. Reduced isoelectric focusing gels were run in urea, and used to assess rFIX heterogeneity and isoform composition. The gamma carboxyglutamic acid content was determined by anion exchange HPLC.

## Example Robustness Study Design

**Purification Process.** rFIX is a family of glycoprotein isoforms secreted by CHO cells, having molecular weights of approximately 55.3 kD, and an isoelectric point within a range of 4.4 - 4.6. Several post-translational modifications occur during the product expression in CHO cells, including N- and O-linked glycosylation, and extensive gamma-carboxylation of the glutamic acid residues located near the N-terminus. Following the production phase in the bioreactor, the CHO cells are removed by filtration, and the cell-free conditioned media is concentrated by ultrafiltration, and then diafiltered to give the desired load conditions for the capture column. The capture step employs anion-exchange chromatography, with adsorption under low ionic strength conditions. The column is washed following the load, and then eluted by a calcium step change which is thought to induce a conformational change in rFIX which causes elution. The elution conditions are restricted to low calcium concentrations to prevent non-specific elution of impurities adsorbed to the resin. Subsequent purification employs orthogonal chromatographic steps to yield the purified bulk drug substance. The capture step yields rFIX of high purity (90 -

99% pure), with good recovery. This robustness study examined the effects of important process variables tested at the ranges described below.

**Process goals.** The goals of the step are to recover sufficient rFIX loaded to the column and yield a process stream of sufficient protein purity to generate a bulk drug substance that meets drug substance purity release specifications after the subsequent purification steps. A minimally acceptable recovery is not easily defined, but a useful definition would limit a worst-case recovery to less than two standard deviations below the average step recovery, based on process-scale data; for this process step, this sets the minimum recovery at approximately 35%.

A minimum purity for the rFIX pool is not specified, since the subsequent chromatographic steps remove residual impurities. Therefore, the protein purity of this in-process stream may not be predictive of the final bulk drug substance purity. The subsequent steps must be performed to assess whether the fraction of protein impurities remaining in the rFIX pool are removed to sufficiently low levels to meet bulk drug substance release specifications.

The study will also determine if there are any significant changes in rFIX composition which would cause product specifications to be exceeded resulting from changes in the tested variables, as measured by the various assays described in the Materials and Methods section.

**Choice of variables and ranges investigated.** The major operations of the step include column equilibration, load, wash, and elution. The variables and test ranges were chosen following the criteria listed above. The rationale for each variable's inclusion, its potential process impact, and the test ranges are given below. A summary of the variables and their test ranges is provided in Table I.

**Load Flowrate.** In order to allow flexibility during the manufacturing of rFIX, it is necessary to define a range for the load flowrate. A fast load flowrate could cause rFIX breakthrough during the load due to mass transfer limitations, especially in the presence of the PVA present in the load from the cell culture medium. The load flowrate was tested at the two levels proposed as control limits for the manufacturing process.

**rFIX Load Mass.** The mass of rFIX loaded per resin volume may influence the column performance, with large challenges leading to rFIX breakthrough and reduced recovery. An operating range was selected well below the dynamic capacity limit of 650 U/ml, which should ensure consistent performance for a range of load masses. The load mass ranged from 160 to 430 U/ml, a 2.7-fold range, and was varied by altering both the load volume and the rFIX concentration in the load by the addition of purified rFIX. The corresponding load masses range from 0.6 - 1.6 mg/ml.

**Load Conductivity and pH.** The load conditions of conductivity and pH, which mediate protein interactions with the anion exchange resin were coupled into a single variable. High pH and low conductivity combine to give the strongest interaction between protein and resin, while conditions of low pH and high conductivity give the weakest interaction. This change in affinity could affect either rFIX recovery, rFIX composition, or the binding of impurities. The pH range was set at 0.2 units above or below the target pH, which is a typical operating range for loads which arise from diafiltered solutions. The conductivity range was based on the established limits for the manufacturing buffer preparation. The loads were prepared at low conductivity and the pH adjusted by titration. Concentrated NaCl stock solution was added for the high conductivity runs.

**Load Volume and Impurity Mass.** The load volume to the column does not vary significantly during manufacture, as the ultrafiltration step delivers a consistent volume to the column. However, by doubling the load volume, the impurity mass is doubled, allowing the impurity mass to be tested as a variable. Higher impurity challenges may result in a reduction in rFIX purity. A two-fold range of load volumes was tested, and the rFIX concentration manipulated by the addition of purified rFIX as described above to give independent variation in load mass and impurity mass.

**Load PVA Concentration.** Due to changes in the bioreactor harvest volume, the choice of a consistent load volume results in varying concentration factors for the ultrafiltration step preceding the capture column. Retention of the PVA present in the media by the UF membrane results in varying polymer concentrations in the load, which affects viscosity, and could potentially affect the dynamic capacity of the column. The PVA level was tested at the typical level resulting from a full harvest, and at 50% of this level. The PVA level was reduced by dilution of the load with buffer lacking PVA.

**Wash Conductivity and Volume.** The wash buffer conductivity affects the adsorbed proteins' affinity for the resin, and the wash volume may influence rFIX breakthrough or impurity removal. These variables were coupled, with high buffer conductivity and long washes giving the greatest loss of adsorbed species, and vice versa. The wash conductivity upper limit was set from process development experiments which found a marked decrease in rFIX recovery at wash conductivities higher than 15.0 mS/cm. The lower limit was derived from manufacturing data, and set at the mean conductivity minus three standard deviations. The wash conductivity was controlled by adding NaCl to the buffer, and the wash volume by programming the LC system accordingly.

**Elution pH and Conductivity.** The elution conductivity and pH may influence rFIX recovery, composition, or impurity level, and were coupled into a single variable, with high pH and high conductivity as the best-case for rFIX recovery, and low pH and low conductivity as the worst-case. A pH limit of plus or minus 0.2 pH unit was applied in order to test the manufacturing control range. The conductivity was set at the mean of the manufacturing experience plus or minus three standard deviations.

**Table I. Tested variables and ranges for fractional factorial study**

| *Variable* | *Lower Limit Tested (-1)* | *Upper Limit Tested (+1)* |
|---|---|---|
| Load Flowrate | 1.6 cm/min | 2.2 cm/min |
| Load Mass | 160 U/ml resin | 430 U/ml resin |
| Load pH and Conductivity | pH 8.0<br>8.4 mS/cm | pH 7.6<br>13.3 mS/cm |
| Load Volume and Impurity Mass | 8.3 CVs | 16.6 CVs |
| Load PVA Concentration | 2.25% | 4.5% |
| Wash Conductivity and Volume | 12.6 mS/cm<br>5 CVs | 15.0 mS/cm<br>7 CVs |
| Elution pH and Conductivity | pH 8.1<br>8.0 mS/cm | pH 7.7<br>11.3 mS/cm |

**Untested Variables.** The untested variables and operations for the chromatographic step include parameters for column equilibration, the column temperature, and conditions for column strippng and regeneration. Column equilibration parameters were not included because the load volume (approximately 8 column volumes) rapidly re-equilibrates the column to the conductivity and pH of the load. Column temperature generally has little influence on ion-exchange chromatography over the narrow control range for this step (2 - 8°C). Wash pH is not expected to influence protein affinity significantly at the low conductivities of the wash. The column strip and regeneration steps occur after elution of rFIX from the column, and are thus considered non-critical, and were subjected to other validation studies.

**Outputs.** rFIX recovery in the product pool is expressed as a percentage of the rFIX units loaded onto the column. The purity of the in-process pool is expressed as a percentage of the total protein in the rFIX pool as measured by the host cell protein ELISA. Other outputs such as rFIX activity in various process fractions (wash, pre- and post-peak pools, regeneration and strip pools), rFIX pool volume, and rFIX pool protein concentration were analyzed to provide supportive information about the process performance. rFIX composition was determined by isoelectric focusing electrophoresis, size exclusion HPLC for detection of aggregates, and analytical Mono Q chromatography to assess gamma carboxyglutamic acid composition of the rFIX.

The primary goal of the step is to achieve sufficient purity in the elution peak to generate acceptable bulk drug substance after the final purification step is completed. This output was evaluated by purifying the rFIX pool from the run that represented the worst case (highest impurity level) challenge of the subsequent purification steps.

**Fractional factorial design and execution.** A total of ten variables were investigated, comprising four single and three pairs of coupled variables. A 16 run Resolution IV study was chosen. This gives unequivocal estimation of all seven main effects, although the 21 two-way interactions are confounded as shown in Table II. Duplicated centerpoints were included, but are not analyzed here.

**Table II. Confounding pattern of the fractional factorial design**

| |
|---|
| Load Flowrate * Load pH/Cond = Load Mass * Elution pH/Cond = Load PVA * Wash Cond/Vol |
| Load Flowrate * Load Vol = Load Mass * Wash Cond/Vol = Load PVA * Elution pH/Cond |
| Load Flowrate * Load Mass = Load pH/ Cond * Elution pH/Cond = Load Vol * Wash pH/Cond |
| Load Flowrate * Load PVA = Load pH/Cond * Wash Cond/Vol = Load Vol * Elution pH/Cond |
| Load Flowrate * Wash Cond/Vol = Load pH/Cond * Load PVA = Load Vol * Load Mass |
| Load Flowrate * Elution pH/Cond = Load pH/Cond * Load Mass = Load Vol * Load PVA |
| Load pH/Cond * Load Vol = Load Mass * Load PVA = Wash Cond/Vol * Elution pH/Cond |

The chosen variables result in eight unique column load solutions, based on high and low combinations of the three load variables (rFIX mass, load volume and load pH /conductivity). Two buffers were prepared for both the wash and elution. Separate programs were written for the execution of the study by the automated chromatographic system, to vary the wash volume as needed. The runs were randomized to avoid systematic errors influencing the analysis; the experimental conditions are given in Table III.

**Data analysis**. Statistical analysis of the data was performed using JMP software, from SAS Institute. All outputs were tested with a family of Box-Cox transforms with main effects models (*3*) to determine if any common transformations decreased the error sum of squares significantly. In only one case did the best transformation alter the form or significance of the model derived from fitting the untransformed output (this is discussed in detail in the section describing rFIX purity below). The model containing all estimable effects was then analyzed to generate a reduced model having only statistically significant terms, which predicts the process output under all conditions, and is thus useful in defining the worst-case. Forward stepwise regression was performed, and the addition of new model terms ended when the next term added was not significant at the 95% confidence level.

**Table III. Experimental conditions and measured outputs**

| | *Load Flowrate* | *Load pH/Cond* | *Load Volume* | *Load Mass* | *Load PVA* | *Wash Cond/Vol* | *Elution pH/Cond* | *Recovery (%)* | *Impurity (%)* |
|---|---|---|---|---|---|---|---|---|---|
| 1 | -1 | -1 | -1 | 1 | -1 | 1 | -1 | 68 | 0.04 |
| 2 | 1 | -1 | 1 | -1 | -1 | -1 | -1 | 62 | 1.47 |
| 3 | -1 | -1 | 1 | 1 | 1 | -1 | -1 | 73 | 0.4 |
| 4 | -1 | -1 | -1 | -1 | 1 | -1 | 1 | 56 | 18 |
| 5 | 1 | 1 | 1 | -1 | 1 | -1 | 1 | 81 | 2.8 |
| 6 | 1 | 1 | 1 | 1 | -1 | 1 | -1 | 57 | 0.04 |
| 7 | 1 | -1 | 1 | 1 | 1 | 1 | 1 | 79 | 1.6 |
| 8 | -1 | 1 | -1 | -1 | -1 | -1 | -1 | 65 | 0.08 |
| 9 | 1 | 1 | -1 | 1 | 1 | -1 | -1 | 80 | 0.04 |
| 10 | -1 | 1 | 1 | -1 | 1 | 1 | -1 | 48 | 0.13 |
| 11 | -1 | 1 | 1 | 1 | -1 | -1 | 1 | 74 | 0.62 |
| 12 | 1 | 1 | -1 | -1 | -1 | 1 | 1 | 77 | 0.46 |
| 13 | 1 | -1 | -1 | -1 | 1 | 1 | -1 | 51 | 0.13 |
| 14 | 1 | -1 | -1 | 1 | -1 | -1 | 1 | 77 | 4.55 |
| 15 | -1 | -1 | 1 | -1 | -1 | 1 | 1 | 86 | 4.61 |
| 16 | -1 | 1 | -1 | 1 | 1 | 1 | 1 | 85 | 0.4 |

## Results

**Product Recovery**. The rFIX recoveries were calculated by the clotting assay, and the results are given in Table III. The histogram in Figure 1 shows the distribution of recovery values. The recovery of the sixteen runs averaged 70 ± 12%, with a range of 48 - 86%. The Pareto plot of the individual factors is shown in Figure 2, and indicate a gradual decrease in significance for the first six variables ( |scaled estimate| ≥ 2.4%), and many insignificant variables ( |scaled estimate| ≤ 1.5%). A stepwise regression analysis identifies a reduced model having only five terms, with no other effects significant at the 95% confidence level. The ANOVA of this model is given in Table IV. Note that each two-way interaction listed in the ANOVA model is indistinguishable from its confounding partners listed in Table II; the two-way interactions included in this ANOVA were chosen for convenience, and do not imply any bias for that pair. Other measures of rFIX recovery (absorbance at 280nm and size exclusion HPLC) also confirm the importance of most of these variables (data not shown), although there are slight differences in the reduced model parameters and the magnitude of the coefficients. The five-parameter reduced model would predict a worst-case recovery of 46%, and a best-case recovery of 94%. This worst-case recovery is well below the average of 70%, but would not be considered a process failure.

**Table IV. ANOVA of the rFIX recovery model for the fractional factorial study**

| Model<br>% Recovery= Int. + $C_1$ (Elution pH/Cond) + $C_2$ (Load pH/Cond * Load Vol) + $C_3$ (Load Mass) + $C_4$ (Load Vol * Load Mass) + $C_5$ (Load Flowrate * Load PVA) | | |
|---|---|---|
| Summary of Fit | adj $R^2$ | 0.88 |
| Parameter Estimates | | |
| Term | Estimate | Prob >\|t\| |
| Intercept | 70 | ≤0.001 |
| $C_1$ (Elution pH/Cond) | 6.9 | ≤0.001 |
| $C_2$ (Load pH/Cond * Load Vol) | -5.9 | ≤0.001 |
| $C_3$ (Load Mass) | 4.2 | 0.008 |
| $C_4$ (Load Vol * Load Mass) | -3.4 | 0.022 |
| $C_5$ (Load Flowrate * Load PVA) | 3.1 | 0.036 |

Further information on the factors affecting rFIX recovery can be gathered from analysis of rFIX activity measurements of in-process pools. Table V lists the factors which affected

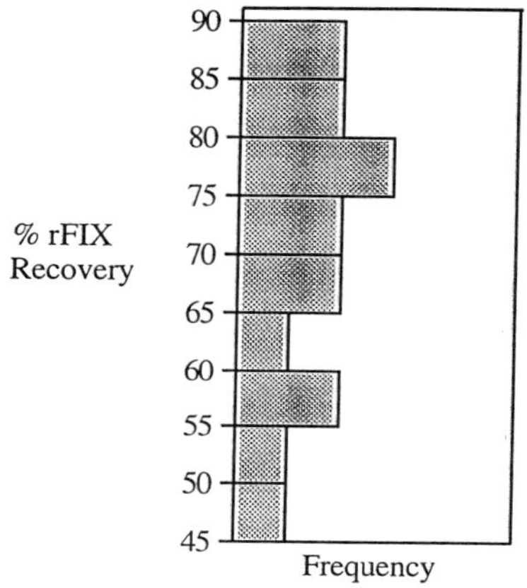


*Figure 1. Histogram of rFIX recovery values for fractional factorial study.*

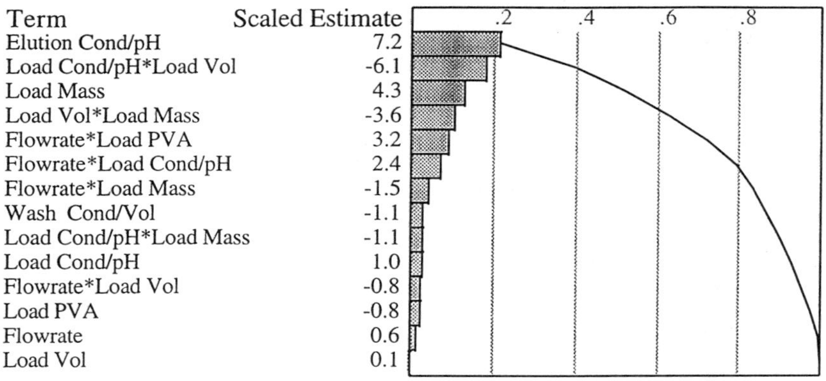


*Figure 2. Pareto plot of factors affecting rFIX recovery for fractional factorial study.*

the rFIX losses to the load eluate, wash eluate, and post-peak and strip fractions (data not shown). The magnitudes for each effect listed are for the full-range of each variable (twice the value of the coefficient of an ANOVA model). The analysis of interim outputs employs smaller models, because some variables have not yet been manipulated (e.g., the load eluate is analyzed with models lacking wash or elution variables, and any interactions containing these terms). The effects in Table V are smaller than the effects of the variables found when the rFIX recovery was analyzed, but this is likely due to the random variation in the assay for highly concentrated samples masking these small effects. The results from the analysis of in-process pool data benefit from the greater statistical confidence of a model having fewer terms and by the larger signal-to-noise ratio for the column eluates than for fractions containing relatively lower activity levels than the peak pool.

**Table V. Factors contributing to rFIX loss in in-process fractions**

| *FRACTION* | *SIGNIFICANT FACTOR* | *MAGNITUDE OF rFIX LOSS* |
|---|---|---|
| Load Eluate | Load pH/Cond | + 2.5% |
| Wash Eluate | Load pH/Cond<br>Wash Cond/Vol | + 2.8%<br>+ 2.2% |
| Post-peak | Elution pH/Cond | - 2.8% |
| Strip | Elution pH/Cond | - 5.6% |

There is no effect of the remaining variables on recovery, suggesting that the process recovery is robust with respect to these variables over the range tested. Some speculations can be made for the mechanisms for the major factors identified in the reduced model. For example, the sensitivity of rFIX recovery to the coupled elution variables is consistent with an increase in the rFIX affinity for the resin during elution causing rFIX to be retained by the resin, and therefore increasing the rFIX levels in the post-peak and strip fractions (see Table V). In addition, the interaction between the load volume and the load pH/conductivity may be explained by the lower than average recovery at high load volumes and high load conductivity.

**Impurity Levels.** The impurity levels of the elution peaks were measured by ELISA, and are given in Table III. The impurity level ranges widely, as indicated in Figure 3. This highly skewed distribution had an average of 2.2 ± 4.5%, with a range of 0.04 - 18.0%. The highest impurity level of 18% was corroborated by SDS-PAGE gel analysis, which showed a marked increase in the intensity of the contaminant bands for this run. The Pareto plot in Figure 4 identifies the two largest factors affecting the impurity levels as the coupled load and elution variables, although the magnitude of their effects is not significantly greater than the remaining variables. Table VI is a summary of the purity

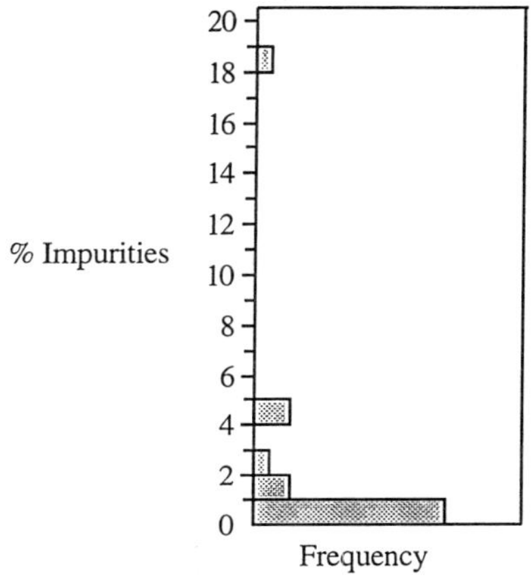


*Figure 3. Histogram of impurity levels for fractional factorial study.*

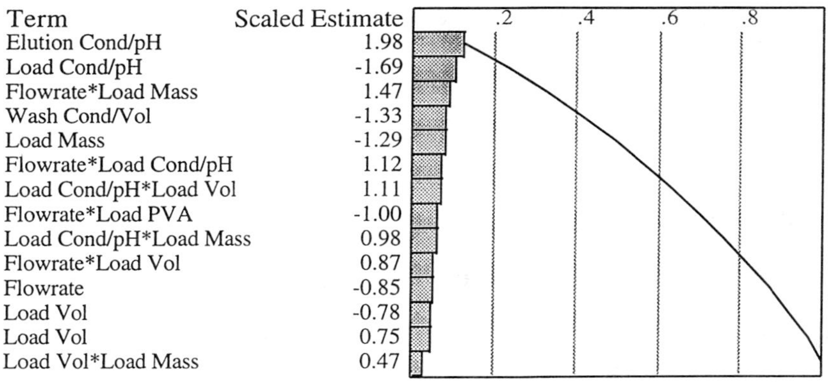


*Figure 4. Pareto plot of factors affecting impurity levels (untransformed data).*

data, grouped by these two variables. The effect of the two-way interaction is clearly seen, as the combination of low load conductivity and high pH and high elution conductivity and low pH causes a significant increase in the impurity level; in fact, the four highest impurity levels were all found under these conditions. This is consistent with the principles of ion-exchange chromatography, as these load conditions will cause more impurities to bind, and these elution conditions will cause more of the bound proteins to elute. However, a stepwise regression analysis failed to identify a model having any significant terms for the raw (untransformed) data.

Because of the clear dependence of the impurity levels on the coupled load and elution variables, as indicated by Table VI, a model containing these two terms was examined. The ANOVA indicated that the model fit poorly (correlation coefficient, r^2, of only 0.38), and none of the coefficients for the two variables was significant at the 95% confidence level. Examination of the residuals from this model show an increase in the residual with the level of impurities, a condition know as heteroscedasticity (*23*), which indicates that a data transform should be tested to stabilize the residuals. Common data transforms including the inverse, logarithm, square root, and inverse square root are all tested by the family of Box-Cox transforms (*3*). The family of Box-Cox transforms was screened using a model having only the main effects, and the sum of squared errors was minimized as a function of lambda (*3*). Figure 5 shows the sum of squared errors (SSE) of the main effects model as a function of lambda, the single adjustable parameter for the Box-Cox transforms. The horizontal line indicates the value of the SSE necessary to have a model of greater than 95% confidence; clearly, the transforms having values of lambda between -0.2 and 0.1 have models fitting the transformed data with greater statistical significance than the untransformed (lambda = 1) data. The screening demonstrated that the log transform (lambda = 0) was the optimum, and the frequency histogram of the natural logarithm of the impurity data is shown in Figure 6. Note that the highly skewed impurity distribution is now more normally distributed, with greater resolution between the runs having lower impurity levels, and a compression of those with higher impurity levels.

**Table VI. Impurity levels as a function of the two pairs of controlling variables**

| Average Product Purity of Four Runs at Each Condition (% of total protein) | | |
|---|---|---|
| | Elution pH/Conductivity (-1) | Elution pH/Conductivity (+1) |
| Load pH/Conductivity (+1) | 0.07 ± 0.04 | 1.07 ± 1.16 |
| Load pH/Conductivity (-1) | 0.51 ± 0.66 | 7.19 ± 7.34 |

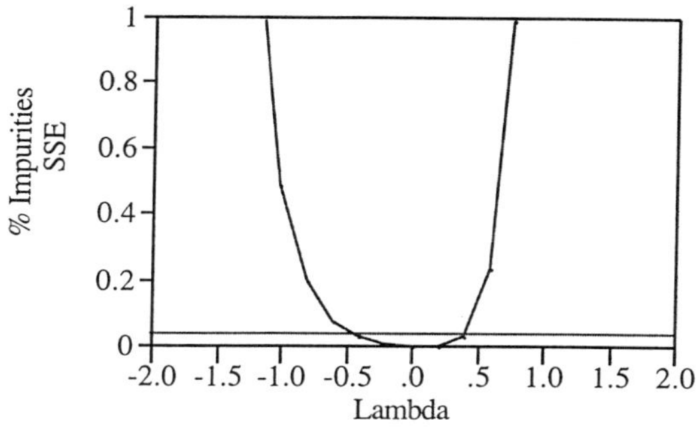


*Figure 5. Box-Cox screening of various transformations of the impurity levels using a main effects model.*

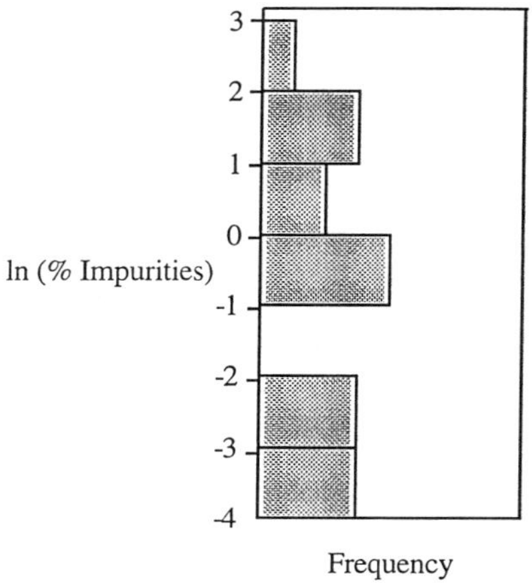


*Figure 6. Histogram of transformed impurity levels.*

The Pareto plot for the transformed data is shown in Figure 7, and the effects of the load and elution variables are now clearly seen. The ANOVA of a reduced model containing these terms is given in Table VII; the effect of performing the transform on the data clearly results in a better model, having an increased correlation coefficient of 0.71, and highly significant coefficients for the two terms (> 99% confidence level). One interpretation of the improved model fit of the transformed data is that the effects of the variables on impurity levels are not linear, but rather follow an exponential dependence. This example illustrates the importance of data transformations when the residuals are not well-behaved.

**Table VII. ANOVA of the impurity model for the transformed data**

| Model ln (% Impurity) = Intercept + $C_1$ (Load pH/Cond) + $C_2$ (Elution pH/Cond) | | |
|---|---|---|
| Summary of Fit | adj $R^2$ | 0.71 |
| Parameter Estimates | | |
| Term | Estimate | Prob >\|t\| |
| Intercept | -0.72 | 0.015 |
| $C_1$ (Load pH/Cond) | -0.8 | 0.009 |
| $C_2$ (Elution pH/Cond) | 1.38 | <0.001 |

When the sensitivity of the impurity level to the coupled elution variables was discovered, a follow-up study was performed to decouple the elution variables. A full-factorial plus centerpoint design was performed, with the elution pH and conductivity set at their tested values from the original 16 run study. The five test conditions and the measured impurity levels are shown in Table VIII.

**Table VIII. Decoupling study conditions and resulting impurity levels**

| *Run #* | *Elution Conductivity (mS/cm)* | *Elution pH* | *Impurity Level (%)* |
|---|---|---|---|
| 1 | 8.1 | 7.7 | 0.7 |
| 2 | 11.5 | 7.7 | 6.4 |
| 3 | 8.1 | 8.1 | 0.2 |
| 4 | 11.5 | 8.1 | 9.7 |
| 5 | 9.3 | 7.9 | 1.2 |

The ANOVA of the log transform of these variables is shown in Table IX, and indicates that the elution conductivity is the dominant variable influencing the impurity level, and that the elution pH does not have a statistically significant effect over this range. Figure 8 shows the effect of the elution conductivity on the impurity levels for these five runs.

**Table IX. ANOVA of the impurity model for the decoupling study**

| Model | ln (% Impurity) = Intercept + $C_1$ (Elution pH) + $C_2$ (Elution Cond) | |
|---|---|---|
| Summary of Fit | adj $R^2$ | 0.86 |
| Parameter Estimates | | |
| Term | Estimate | Prob >\|t\| |
| Intercept | 0.56 | 0.17 |
| $C_1$ (Elution pH) | -0.21 | 0.553 |
| $C_2$ (Elution Cond) | 1.52 | 0.035 |

Recognizing that the higher elution conductivity level caused significant changes to the levels of impurities in the in-process pools, a decreased range was proposed for this variable. A review of the manufacturing data obtained subsequent to the fractional factorial study design and execution indicated that the buffer conductivities were in fact very tightly controlled (a mean and relative standard deviation of 10.2 ± 0.7 mS/cm for 46 different lots), and so a reduction in the upper limit for conductivity from 11.5 to 11.0 mS/cm was adopted into the manufacturing formulation record limits. Based on a similar analysis of the manufacturing data for the diafiltration buffer of the ultrafiltration step prior to the capture step, an additional change was made to the load conductivity limit, raising the lower conductivity limit from 8.3 to 9.0 mS/cm. These two changes to the process limits were then tested by performing triplicate runs, combining the elution pools, and processing them through the remaining purification steps. The in-process impurity levels of the triplicate runs averaged 8.0 ± 0.2%, in agreement with the prediction of a model derived from the transformed data, containing the terms for the coupled load conditions and the elution conductivity (9.2%). The bulk drug substance purified from these three runs was found to meet the specifications for rFIX purity, indicating that the new elution conductivity limit provides a reasonable assurance that this processing step is capable of generating a rFIX pool which will yield a product meeting all purity criteria following subsequent purification steps.

**Product Composition.** Several aspects of rFIX composition were also tested for the in-process pools generated in the original 16-run study. These tests included measures of high-molecular weight aggregates by size exclusion HPLC, rFIX heterogeneity by isoelectric focusing, and the degree of gamma carboxylated glutamic acid modification.

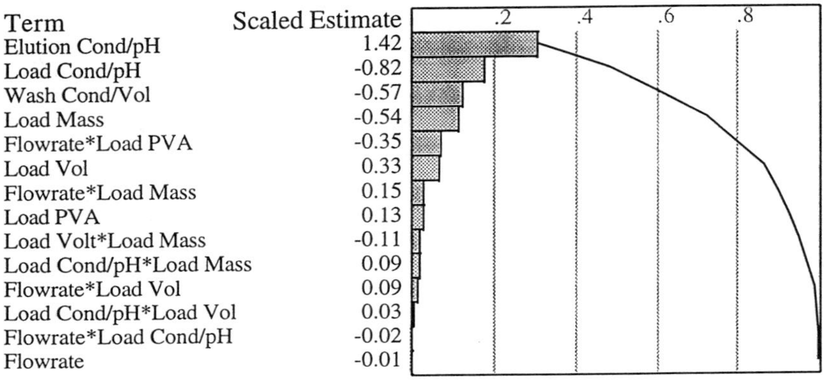


*Figure 7. Pareto plot of factors affecting impurity levels (transformed data).*

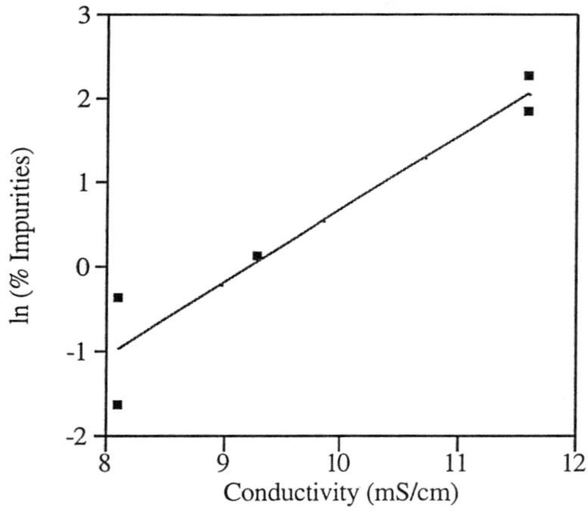


*Figure 8. Effect of elution conductivity on impurity levels from decoupling study.*

In all cases, there was either no significant reduced model, or the reduced model had no process significance, as the worst-case prediction was well within the allowable range. This confirms that the process is capable of producing material having adequate product quality as measured by these assays, for the variables examined over the ranges tested.

**Discussion**

The recovery of rFIX across this chromatographic step was modestly influenced by several variables over the ranges tested. The variables were found to interact (that is, the effect of one variable depended upon the level of another variable), which emphasizes the utility of an experimental methodology which detects interactions between variables, something that one-at-a-time testing cannot estimate. The worst-case prediction of the reduced model of the rFIX recovery was acceptable. This finding confirms that process variation within the ranges tested for the ten variables investigated will produce an acceptable yield of rFIX. Because of the narrow ranges of the variables tested, and the fact that the rFIX recoveries of the centerpoint runs are also close to the mean recovery (data not shown), this suggests that intermediate levels of the influential variables will not cause a process failure as measured by rFIX recovery. Future studies could examine increases in recovery by optimizing some of the variables found to influence recovery. For example, the load pH and conductivity could be decoupled to determine which is the stronger influence on rFIX recovery. The control range of the important variable determined by the decoupling could then be reduced or re-centered to maximize the product recovery. The problem of competing or mutually exclusive optima between product recovery and product purity could arise if elution conductivity were the dominant variable (high elution conductivity may both increase rFIX recovery and raise the impurity levels in the peak pool). Such a situation would require a careful assessment of which conductivity range would strike the appropriate balance between the two outputs.

The impurity levels of the in-process peak pool generated by this step were a strong function of two pairs of coupled variables. An analysis of the raw data failed to yield a statistically significant model; a transformation was required in order to allow interpretation of the data. The transformation indicates that there is an exponential dependence of the impurity levels on the load and elution variables of pH and conductivity. A subsequent decoupling study found the elution conductivity to have the greatest influence on the impurity level. After combining this data with manufacturing information on buffer preparation, reduced ranges for elution buffer conductivities were proposed and tested at prospective scale. The impurity levels were reduced from the maximum seen in the 16-run fractional factorial study, in excellent agreement with the reduced statistical model. It is expected that this process change will reduce the level and variability of the in-process pool impurities. An acceptable purity was achieved for rFIX bulk drug substance derived from the worst-case conditions of the elution conductivity. This example illustrates how the implementation of a process modification based on the modeling of data generated by a robustness study can be performed. This tuning of the process parameters arises from an increased understanding of the controlling factors for the process; in this case, the sequential fractional factorial studies allowed the determination of which of two narrow variable ranges controlled in-process impurity levels.

Finally, various product quality attributes were found to be acceptable for all conditions tested. While significant models were found for some composition outputs, the magnitude of the variables' effects were so small as to not have any process significance, as the variability in the bulk drug substance was well within the product specification range. It is also possible that the small but statistically significant effects seen are due to the choice of an ANOVA model with effects significant at the 95% confidence level, which will generate a model containing false positive effects with an average frequency of 1 in 20.

Thus, this chromatographic purification step used for the manufacture of rFIX may be considered to be robust with respect to all ten variables tested and their numerous combinations. Because the variables examined represent the most probable sources of process variation, this study provides a high degree of assurance that the step will perform adequately despite normal variations in feedstream, buffer composition, and column operation.

The evaluation of process robustness by fractional factorial methodologies requires a significant outlay of time and resources. The decision to perform such studies on all process steps can greatly increase the burden on a validation program. An alternative strategy would be to perform such studies only for critical processing steps, for which robust operation is paramount.

## Acknowledgments

The authors would like to acknowledge helpful discussions with Jeff Deetz and Barry Foster.

## Literature Cited

(*1*) Food and Drug Administration. *Guidelines on General Principles of Process Validation,* 1987.

(*2*) Food and Drug Administration. *Text on Analytical Validation (ICH Guidelines on Validation of Analytical Procedures: Definition and Terminology),* 1995.

(*3*) Box, G.E.P.; Hunter, W.G.; Hunter, J.S. *Statistics for Experimenters*; J. Wiley: New York, NY, 1978.

(*4*) Daniel, C. *Applications of Statistics to Industrial Experimentation*; J. Wiley: New York, NY, 1976.

(*5*) Haaland, P.D.; *Experimental Design in Biotechnology*; Marcel Dekker: New York, NY, 1989.

(*6*) Mason, R.L.; Gunst, R.F.; Hess, J.L. *Statistical Design and Analysis of Experiments*; J. Wiley: New York, NY, 1989.

(*7*) Montgomery, D.C. *Design and Analysis of Experiments*; J. Wiley: New York, NY, 1991.

(*8*) Cruz, P.M.; Christen, P.; Farres, A. *J. Ferm. Bioeng.* **1993**, *76*, pp. 94-97.

(*9*) Emborg, C.; Jepsen, P.K.; Biedermann, K. *Biotechnol. Bioeng.* **1989**, *33*, pp. 1393-1399.

(*10*) Greasham, R.; Inamine, E. In *Manual of Industrial Microbiology and Biotechnology;* Demain, A.L.; Solomon, N.A., Ed.; American Society of Microbiology: Washington, D.C., 1986.

(*11*) Greasham, R. In *Biotechnology;* Stephanopolous, G. Ed.; VCH Publishers: Weinheim, Vol 3, Bioprocessing.

(*12*) Serralheiro, M.L.M; Cabral, J.M.S. *Biotechnol. Bioeng.* **1994**, *43*, pp. 1031-1042.

(*13*) Vindevogel, J.; Sandra, P. *Anal. Chem.*, **1991**, *63*, pp. 1530-1536.

(*14*) Weatherell, C.J.; Lai, E.P.C. *J. Chrom.,* **1994**, *669*, pp. 31-44.

(*15*) Berggrund, A.; Drevin, I.; Knuuttila, K.-G.; Wardhammar, J.; Johansson, B.-L. *Proc. Biochem.*, **1994**, *29*, pp. 455-463.

(*16*) Gaertner, J.G.; Dhurjati, P. *Biotechnol. Prog.*, *9* pp. 298-308.

(*17*) Gaertner, J.G.; Dhurjati, P. *Biotechnol. Prog.*, *9* pp. 309-316.

(*18*) Kelley, B.D.; Jennings, P; Wright R.; Briasco, C. *BioPharm*, **1997**, *10 (10)*. pp. 36-47.

(*19*) Jobson, J.D.; *Applied Multivariate Data Analyis*; Springer-Verlag: New York, NY,1992.

(*20*) Barry, A.R.; Chojnacki, R. *BioPharm* **1994**, *7*, pp. 43-47.

(*21*) Kuwahara, S.S.; Chuan, J.H. In *Separations Technology: Pharmaceutical and Biotechnology Applications*, Olson, W.P., Ed.; Interpharm Press: Buffalo Grove, IL, 1996; pp. 427 - 451.

(*22*) Sofer, G.; Hagel, L. *Handbook of Process Chromatography: A Guidϵ to Optimization, Scale-up, and Validation*; Academic Press: San Diego, CA, 1997.

(*23*) Zar, J. H. *Biostatistical Analysis*; Prentice-Hall: Upper Saddle River, NJ, 1996.

## Chapter 9

# Virus Removal and Inactivation

## A Decade of Validation Studies: Critical Evaluation of the Data Set

**Joachim K. Walter, Franz Nothelfer, and William Werz**

**Boehringer Ingelheim Pharma Deutschland KG, D-88397 Biberach, Germany**

Recombinant DNA technology as well as hybridoma technology provide an outstanding opportunity for the preparation of complex human proteins with significant impact on replacement therapy. Acting as living production facilities, the microorganisms and animal cells used for the manufacturing of proteins are subject to biological deviations and are influenced by the environment.

With respect to mammalian cell culture, the cell's susceptibility for potential viral infections is of concern regarding the drug safety. Measures for the removal of potential viral contaminants during the downstream processing would be more reliable, if only cell cultures essentially free of endogenous virus are employed for the expression of the desired protein product; however, potential viral contamination may be induced adventitiously by application of biological raw materials to the production process. In addition, a failure in proper containment e.g. in venting and aeration of the fermentation culture may definitely lead to a viral infection. Especially in the case of viruses of unknown species, it is theoretically and practically impossible to monitor the inactivation or removal of such species. As they are expected to have unknown but potentially harmful biological effects, specific assays can not be devised to monitor their presence and/or behavior during the manufacture of a pharmaceutical protein.

Hence, a number of "next best" preventative measures have to be established around the manufacturing process. All have the common goal to reduce the probability of a potential viral contamination. These measures include tests for a great number of specific viruses potentially present in the designated Master Cell Bank (MCB) and define specific tests to detect potential adventitious virus infections during the production process. Nevertheless, the capability and the capacity to reduce potential viral contaminants has to be demonstrated for individual unit operations of the downstream processing. To this end however, model viruses must be selected which represent the whole range of virus species. Table I lists the model viruses typically used in validation studies. Criteria for the selection of such viruses are:

**Table I: Model Viruses for Validation Studies (adapted from [7, 8])**

| Virus | Familiy | Natural Host | Genome | Envelope | Size | Shape | Resistance to Physicochemical Treatment |
|---|---|---|---|---|---|---|---|
| Vesicular stomatitis virus | Rhabdo | Equine Bovine | RNA | + | 70 x 175 nm | Bullet shape | Low |
| Parainfluenza virus | Paramyxo | Various | RNA | + | 100 - 200 nm | Pleo/Spherical | Low |
| Human immuno-deficiency virus | Retro | Man | RNA | + | 80 - 100 nm | Spherical | Low |
| Murine leukaemia virus (MuLV) | Retro | Mouse | RNA | + | 80 - 110 nm | Spherical | Low |
| Poliovirus Sabin type I | Picorna | Man | RNA | - | 25 - 30 nm | Icosahedral | Medium |
| Encephalomyo-carditis (EMC) | Picorna | Mouse | RNA | - | 25 - 30 nm | Icosahedral | Medium |
| Reovirus 3 | Reo | Various | RNA | - | 60 - 80 nm | Spherical | Medium |
| SV 40 | Papova | Monkey | DNA | - | 40 - 50 nm | Icosahedral | Very High |
| Parvoviruses (canine, porcine) | Parvo | Canine Porcine | DNA | - | 18 - 24 nm | Icosahedral | Very High |

- size of the viral particle
- enveloped / non-enveloped virus
- genome structure: DNA / RNA
- strandedness of the virus genome
- resistance to inactivation methods

With regard to the feasibility of validation work on virus inactivation / removal, criteria also include:

- relevance of the virus species for the producer cells
- achievable high virus titer
- high sensitivity of detection
- ease of detection

Virus clearance during downstream processing is performed by methodologies for virus inactivation and virus removal. Regarding virus removal, two principal unit operations of the downstream process may contribute: filtration and chromatography (*1- 4*).

This paper presents data from numerous validation studies on the inactivation and removal of virus during the downstream processing. This set of data has been collected, compared and evaluated to rank such unit operations regarding their impact, importance and reliability for the drug safety of mammalian cell culture derived pharmaceutical proteins.

**Validation of Virus Removal and Inactivation**

The capability and capacity of individual unit operations in the downstream processing regarding virus removal and inactivation needs to be validated according to the very same concepts and criteria established for process validation in general (*5-8*). With rare exceptions such as microwave-induced High Temperature Short Time (HTST)-Heat Treatment the scale of the respective validation equipment might be different from the manufacturing scale: typically the validation needs to be performed using the equipment at least of identical type and scale as intended for the production; validation of viral clearance however has found acceptance as an exception to this rule: the contamination of equipment for the production with infectious virus as a spike would impair such equipment to an irresponsible degree. In addition, the availability of the required amount of respective virus at high titer for a production scale of several thousand liters is not feasible. Accordingly, the design of the downstream processing should consider some basic requirements for the validation of viral clearance:

- Every process step should be transferable to the required scale.
- All process equipment that is not disposable has to be at least sanitizable.
- The design of all process components must permit the necessary validation procedures.

## Unit Operations for Viral Clearance

Data for viral clearance have been obtained from a number of downstream processes for various unit operations and modes of operation:

- Chromatography
  - Affinity
  - Hydrophobic Interaction
  - Cation Exchange
  - Anion Exchange
  - Gel permeation
- Filtration
  - Nanofiltration
  - Ultrafiltration
- Inactivation
  - Acid
  - Urea
  - Convective Heat
  - Microwave Heat

## Chromatography Data

Any chromatographic separation which is claimed for viral clearance must be validated first with respect to the down-scaled chromatographic process, before it is applied to validation experiments:

- Proof of product quality before and after the chromatographic separation indicates the process performance.
- Validity and reliability of the chromatographic process is demonstrated by the reproducibility and comparability of process data derived from production scale and down-scale. Typical process parameters to be checked are:

| | |
|---|---|
| - yield | - linear flow rate |
| - resolution | - pH-value |
| - selectivity | - conductivity |
| - capacity | - system pressure |
| - profiles | - sample load respective matrix volume |

## Conclusions Chromatography

The results in Table II, which were obtained from a number of different chromatographic processes, indicate that - even comparable modes of chromatography contribute to virus removal to a different and in inconsistent extent. Despite the fact that an individual chromatographic separation is highly reproducible, the data clearly show that chromatography works as it is designed for: it separates proteins by

**Table II: Viral Clearance in Chromatographic Separations**

| Type | Protein | MuLV | PI-3 | Reo3 | SV-40 | Process Buffer | |
|---|---|---|---|---|---|---|---|
| Affinity | | | | | | | |
| Prot A | 150 Kd | 4.0 | 3.2 | <1.0 | 2.1 | L: pH 9.0<br>E: pH 3.0 | 190mS<br>38mS |
| Prot G | 150 Kd | 6.5 | 7.0 | 2.0 | 3.3 | L: pH 6.25<br>E: pH 3.4 | 4.5mS<br>1mS |
| Amino Acid | 60 Kd | >6.3 | 6.5 | 1.1 | 4.0 | L: pH 6.0<br>E: pH 7.5 | 26mS<br>15mS |
| Hydrophobic Interaction | | | | | | | |
| Pyridyl | 46 Kd | 3.6 | 1.7 | >6.8 | 5.5 | L: pH 8.0<br>E: pH 5.5 | 115mS<br>10mS |
| Octyl | 80 Kd | 5.0 | <1.0 | 1.7 | <1.0 | FT: pH 5.8 | 135mS |
| Cation Exchange | | | | | | | |
| S Sepharose FF | 150 Kd | >5.9 | >6.3 | 2.8 | 1.1 | L: pH 6.0<br>E: pH 7.8 | 6mS<br>20mS |
| | 46 Kd | >4.0 | 1.2 | 2.8 | 5.9 | L: pH 6.3<br>E: pH 6.5 | 6mS<br>9mS |
| SP Sepharose FF | 80 Kd | >6.3 | 6.5 | 1.1 | 4.0 | FT: pH5.0 | 5mS |
| | 60 Kd | 3.6 | 1.7 | >6.8 | 5.5 | L: pH 6.8<br>E: pH 7.5 | 8.5mS<br>40mS |
| Anion Exchange | | | | | | | |
| TMAE Fractogel EMD | 150 Kd | >6.3 | 3.0 | 5.3 | 3.5 | L: pH 7.5<br>E: pH 7.5 | 5mS<br>12mS |
| | 80 Kd | >6.7 | 5.3 | 6.1 | >6.7 | L: pH 8.5<br>E: pH 7.5 | 2mS<br>15mS |
| DEAE Sepharose FF | 150 Kd | 5.0 | 6.0 | 6.0 | 5.8 | L: pH 8.8<br>E: pH 8.5 | 4mS<br>8.5mS |
| | 150 Kd | 3.6 | 1.7 | >6.8 | 5.5 | L: pH 8.5<br>E: pH 8.5 | 6mS<br>16mS |
| DE 52 | 60 Kd | 5.0 | <1.0 | 1.7 | <1.0 | FT: pH 7.5 | 15mS |
| Q Sepharose FF | 150 Kd | >5.7 | >5.7 | >6.7 | >6.9 | FT: pH 8.0 | 6mS |
| | 80 Kd | >6.4 | 1.5 | 6.1 | 2.7 | FT: pH 7.5 | 11mS |
| | 60 Kd | 7.0 | 1.3 | >7.7 | n.d. | L: pH 7.2<br>E: pH 7.5 | 6mS<br>40mS |
| | 46 Kd | >3.8 | <1.0 | 5.2 | 5.6 | FT: pH 8.0 | 9mS |

L= Load, E= Elution, FT= Flowtrough

interaction - and does not differentiate at all whether these proteins are product or virus related. As an exception anion exchange chromatography was found to remove virus at basic pH and very low conductivity. Viral clearance on TMAE Fractogel EMD seems to be more consistent than on other anion exchange matrices (see PI-3 data); with TMAE Fractogel, the functional ion exchanger groups are located on long tentacle-type spacers which limit sterical hindrance thus offering a multi-point attachment even to supramolecular structures.

The sensitive murine Leukemia virus as a typical model for retroviruses was removed significantly in all investigated cases. Chromatography is of high value if the virus of concern is identified, and hence can be used for the validation work. Re-use and lifetime of the chromatographic matrix needs to be validated, but is usually of no concern, as the performance of the respective matrix can easily be evaluated by analytical testing. Typically an expensive virus challenge of used matrix is requested, but the accuracy of virus titration as a biological assay is limited, whereas a physico-chemical analysis is by magnitudes more sensitive and hence would be most appropriate.

**Filtration Data**

Ultrafiltration as well as nanofiltration are most feasible for the mechanical removal of viral particles. Evaluation of respective membranes in a down-scaled configuration includes:

- Proof of product quality before and after the filtration procedure.
- Validity and reliability of the filtration process is demonstrated by the reproducibility and comparability of the relevant process parameters applied to production scale and down-scale:
  - inlet pressure
  - outlet pressure
  - transmembrane pressure
  - yield
  - ratio of flow rate: permeate / retentate
  - product load respective membrane area

**Conclusions Filtration**

The investigations of both ultrafiltration and nanofiltration clearly demonstrate that sieving and discrimination of viral particles is consistent and reliable (see Table III). However, the thorough operation of the filtration, namely the tangential flow mode of ultrafiltration, must take care and control of the boundary layer formation; this is decisive for the validity of such filtration process. Ultrafiltration with a nominal cut-off of 100Kd and 200 Kd clear effectively small viruses such as SV-40. The triple-layer membrane of DV50 has an effective pore size of about 35 nm compared to about 70 - 80 nm for Ultipor 40 and Supor 30 and clears SV-40 as well. The operational advantage of cartridge type nanofilters lays evidently in their easy implementation as an in-line device; the implementation into an existing process is highly feasible.

**Table III: Viral Clearance by Filtration**

| Filter | Protein | MuLV | PI-3 | Reo3 | SV-40 | Process Buffer | |
|---|---|---|---|---|---|---|---|
| Ultrafiltration | | | | | | | |
| Omega 100 Kd | 50 Kd | >3.8 | >3.1 | >3.9 | >3.3 | pH 7.1 | 14mS |
| | 46 Kd | >3.8 | >3.0 | >3.1 | >4.3 | pH 5.5 | 22mS |
| Omega 200 Kd | 80 Kd | 4.2 | 3.5 | 5.4 | 2.9 | pH 5.0 | 5mS |
| | 60 Kd | n.d. | >6.9 | >7.2 | >5.5 | pH 7.2 | 6mS |
| Omega 300 Kd | 150 Kd | 4.0 | 5.2 | 2.6 | <1.0 | pH 8.5 | 9mS |
| | 60 Kd | 4.4 | 3.3 | 3.5 | <1.0 | pH 7.5 | 15mS |
| Nanofiltration | | | | | | | |
| Ultipor 40 | 150 Kd | 4.9 | 3.7 | <1.0 | <1.0 | | |
| | 65 Kd | >6.3 | >6.6 | 2.9 | <1.0 | | |
| | 46 Kd | >3.6 | >4.4 | <1.0 | <1.0 | | |
| | 80 Kd | 4.9 | 4.8 | 2.9 | <1.0 | | |
| Supor 30 | 60 Kd | >5.2 | 7.5 | <1.0 | <1.0 | | |
| DV 50 | 60 Kd | >5.9 | >7.1 | >7.5 | >5.8 | | |

## Inactivation Data

Inactivation of virus is definitely the superior methodology compared to removal: the exact and careful determination of the inactivation kinetics allow for a highly reliable and precise performance of the inactivation process and its validation. In general, non-invasive physical methodologies, such as heating, are preferred over invasive methodologies using chemical agents, which have to be removed and require additional expensive analytical monitoring.

The inactivation procedure should cover a broad range of virus species. Some popular inactivation processes like the solvent-detergent treatment or acid treatment are solely feasible for enveloped viruses, hence the inactivation of small non-enveloped viruses, such as SV-40, Polio and highly resistant porcine parvovirus (PPV), remains most challenging. The introduction of microwave-induced HTST-treatment offers for the first time a substantial inactivation of small non-enveloped viruses while fully maintaining the integrity of the protein product. Under appropriate buffer conditions, rtPA could be processed up to 140 °C at 10 mg $ml^{-1}$. A humanized monoclonal antibody was successfully processed at a 2000 L scale at 90 °C and 7 mg $ml^{-1}$ (*9*). As with the removal methods (chromatography and filtration), the evaluation of the down-scale of the inactivation procedures is crucial:

- Proof of product quality before and after the inactivation procedure.
- Validation and reliability of the inactivation procedure is demonstrated by the reproducibility and comparability of the process performance:
  - tank / container geometry
  - mixing efficacy and time
  - incubation time
  - temperature
  - physico-chemical parameter (pH, conductivity, concentration)

With respect to the validation concept, the microwave-induced HTST heating features the unique opportunity to spike and re-collect a virus sample of a volume as low as 20 - 30 ml into the fluid pathway using a designed sample applicator under operational conditions for the manufacturing process (flowrate 35 - 80 L $h^{-1}$, temperature 60 - 165 °C) at full scale (*2*); the complete pathway is disposable, hence offering an extraordinary validation opportunity as well as a multi-product use and avoiding any potential cross contamination.

## Conclusions Inactivation

The data on inactivation procedures using acid, a chaotropic salt or heat verify the expectation that known resistant viruses are difficult to inactivate under conditions which allow for processing of labile proteins (Table IV). In general, the non-invasive physical method of heating was demonstrated to be superior to invasive methods using acid or urea treatment:

**Table IV: Virus Inactivation**

| Treatment | Protein | MuLV | PI-3 | Reo3 | SV-40 | Process Buffer | |
|---|---|---|---|---|---|---|---|
| Acid | | | | | | | |
| pH 4.0 | 60 Kd | 5.0 | 4.2 | <1.0 | <1.0 | pH 7.1 | 14mS |
| pH 4.0 | 80 Kd | 5.1 | 1.2 | <1.0 | 1.1 | pH 5.5 | 22mS |
| pH 3.9 | 60 Kd | 4.2 | <1.0 | <1.0 | <1.0 | pH 5.0 | 5mS |
| pH 3.5 | 60 Kd | >6.0 | 6.5 | <1.0 | <1.0 | pH 7.2 | 6mS |
| pH 3.0 | 46 Kd | >3.0 | >3.8 | <1.0 | <1.0 | pH 8.5 | 9mS |
| Urea | | | | | | | |
| 3.0 M | 60 Kd | >5.2 | 5.3 | 2.6 | n.d. | pH 7.1 | 14mS |
| 4.0 M | 60 Kd | 4.8 | 5.2 | 4.7 | <1.0 | pH 5.5 | 22mS |
| Convective Heat | | | | | | | |
| 60 °C, 10 h | 65 Kd | 3.8 | 3.1 | 3.9 | 3.3 | pH 7.1 | 14 mS |

Microwave Heat

The inactivation profiles of microwave generated heating for different viruses show:

- complete inactivation of enveloped and large non-enveloped viruses (HIV, IBR, EMC, VSV, RSV, Reo 3) at temperature >75 °C
- non-enveloped virus (SV-40) is inactivated by 3.2 log at 83 °C, 4.9 log at 86 °C, 5.5 log at 89 °C and >6.45 log at >92 °C
- small non-enveloped virus (PPV) is inactivated by 1.2 log at 93 °C, 2.0 log at 95 °C, 4.5 log at 97 °C and >6.5 log at > 99 °C

- Acid treatment at pH > 3 does not inactivate non-enveloped viruses, enveloped viruses are not inactivated in a reliable manner.
- Urea treatment does not inactivate small non-enveloped viruses, large non-enveloped viruses (Reo3) are inactivated to some extend.
- Heat inactivation contributes significantly to the inactivation of all investigated viruses. Microwave generated heat was proved to be superior to convective heat with respect to efficacy and process time; even small non-enveloped viruses are fully inactivated (SV40) or inactivated to a high extent (PPV) at temperatures feasible for the processing of pharmaceutical proteins.

## Summary

The experience on virus removal and inactivation during the last decade leads to a distinctive design of the downstream processing of pharmaceutical proteins derived from mammalian cell culture. Generally, a dedicated virus removal / inactivation block is located midstream of the process as shown in the table below:

| Downstream Process | Viral Clearance | Fluid Volume % | Product Concentration |
|---|---|---|---|
| Cell-free Culture Fluid | | 100 | 0.2 mg/ml |
| ⇓ Capture | | 100 - 10 | 0.2 - 2 mg/ml |
| ⇓ Midstream | Virus Inactivation Filtration Chromatography | 10 - 2 | 2 - 10 mg/ml |
| ⇓ Polishing | | 1 - 0.2 | 20 - 100 mg/ml |
| ⇓ Filling | Total Viral Clearance: 12 - 15 log | 20 - 0.2 | 1 - 100 mg/ml |

The data on virus removal and inactivation and their evaluation support a ranking of appropriate methodologies. First choice is the application of reliable methods for the required minimum of viral clearance regarding drug safety. The additional application and validation of chromatographic unit operations within the downstream process, which exhibit a potential for significant clearance of some virus species, may be highly supportive in order to expand the overall drug safety:

- **Rely on inactivation and filtration methods**
  - Ultratherm (Charm BioEngineering Inc, Malden, MA, USA):
    highly effective heat inactivation with unique validation features.
  - DV50 Nanofilter (Pall Corporation, New York, USA):
    excellent virus removal also for small viruses, preferred over ultrafiltration due to its ease of use.

- **Chromatography data supportive only**
  - TMAE Fractogel EMD (E. Merck, Darmstadt, Germany):
    effective for a broad range of viruses; suggests a more consistent viral clearance compared to other anion exchange matrices or modes of chromatography

## References


1. Berthold, W., Walter, J.K., Protein Purification: Aspects of Processes for Pharmaceutical Products, *Biologicals*, **1994**, vol 22, pp. 135-150.
2. Walter, J.K., Werz, W., Berthold, W., in: *Viral Safety and Evaluation of Viral Clearance from Biopharmaceutical Products*, eds. Brown, F., Lubiniecki, A.S., Dev Biol Stand, Karger, Basel, 1996, vol 88, pp. 99-108.

3. Berthold, W., Walter, J.K., Werz, W., Experimental approaches to guarantee minimal risk of potential virus in purified monoclonal antibodies, *Cytotechnology*, **1992**, vol 9, pp. 189-201.
4. Walter, J.K., Werz, W., Berthold, W., Virus Removal and Inactivation, *BFE*, **1992**, vol 9, no 9, pp. 560-564.
5. Walter, J.K., Allgaier, H., Validation of Downstream Process, in: *Mammalian Cell Biotechnology in Protein Production*, eds. Hauser, H., Wagner, R., Walter de Gruyter: Berlin, New York, 1997, pp. 453-482.
6. U.S. Department of Health and Human Services, Center for Biologics Evaluation and Research, Points to Consider in the Manufacture and Testing of Monoclonal Antibody Products for Human Use, FDA, Rockville, MD, USA, 1997.
7. ICH Q5A, Viral Safety Evaluation of Biotechnology Products Derived from Cell Lines of Human or Animal Origin, 1997.
8. CPMP, Virus Validation Studies: Design, Contribution and Interpretation of Studies Validating the Inactivation and Removal of Viruses, CPMP/268/95, 1996.
9. Walter, J.K., Winter, H., Striffler, U., Microwave Heat Inactivation of Virus: Resistance of Pharmaceutical Proteins Exposed to High Temperature Short Time Heat Treatment, IBC's Symposium on Viral Clearance, Philadelphia, PA, USA, Oct 27-28, 1997.

## Chapter 10

# Reuse Validation of an Anion Exchange Chromatography Step for Purification of Clinical-Grade Ciliary Neurotrophic Factor

**Somesh Nigam, Gary Ruezinsky, and James Dugger**

**Regeneron Pharmaceuticals Inc., 777 Old Saw Mill River Road, Tarrytown, NY 10591**

The economical manufacture of biological therapeutics requires repeated use of chromatography column resins during the purification process as a pre-requisite to overall cost reduction. The validation of resin reuse is critical for establishing that the columns would perform reproducibly without any significant deterioration in purification ability or resin integrity up to the maximum number of specified reuse cycles. In this article we present a case study for performing scaled-down column reuse studies involving an anion exchange chromatography step utilized for the purification of clinical-grade human Ciliary Neurotrophic Factor (rHCNTF). The purification parameters examined include elution profiles, product yield/quality, clearance of DNA, endotoxin and *E. coli* proteins and the level of leachables generated from resin matrix breakdown.

As with any production-level purification, the ultimate objective is to reduce unit operation costs, thereby reducing overall manufacturing costs. One of the most obvious and effective means of accomplishing this is the repeated use of chromatography resins, which minimizes material and labor costs during purification (*1-3*). Purification of biotherapeutic products usually requires a sequence of multiple chromatography steps in order to separate the desired product from a heterogeneous mix of contaminants (*1,3,4*). The more complex a purification strategy becomes, the more susceptible it is to minor problems in each individual column's performance. For example, minor amounts of residual impurities remaining on a resin from a previous purification may increase with time and cause severe column performance problems (*1,3*). Since each successive column's performance is dependent upon that of the column preceding it, it is imperative to assure that each column step performs within a well-defined window that determines column performance and in-process product quality over its lifetime(*1-3*). Finally, it is important to consider the potential contribution of the column resin/matrix to background contamination in the form of leachables (*1-3, 5*).

The approach to column chromatography validation has been described in several FDA guidelines and "Points to Consider" documents" as well as in monographs published by industry groups *(1-4 )*. In this article we present a case study involving the development of a validation package for supporting the reuse of an anion exchange column (DEAE Sepharose) utilized for the purification of Regeneron's recombinant human Ciliary Neurotrophic Factor (rHCNTF).

Reuse validation of column performance, depends on the particular requirements of the column or unit operation in question and ability to assemble a system which closely reflects the manufacturing operation. Column requirements arise via a close inspection of the physical and chemical properties of the protein of interest in comparison with the impurities contributed by the host cell. In the case of bacterial cells, these may consist of lipoproteins, lipopolysacharides, proteins, nucleic acids and combinations thereof (*1,3,4*). Monitoring a resin's ability to consistently remove these contaminants from known starting material will demonstrate its suitability for repeated use. A small-scale representation of a manufacturing step involves careful consideration of the nature of the separation and preservation of key physical parameters such as resin bed height and linear flow rate. Reuse run data can then be scrutinized to establish consistency of key column performance parameters relating to chromatographic reproducibility and product quality (*1-4,6*).

Recombinant rHCNTF is expressed in relatively large quantities (approximately 15% of the total protein) in a non-native inclusion body bound form. Initial protein recovery involves a number of steps including extraction, refolding and diafiltration, which remove much of the extraneous host cell contaminants, precede sequential chromatography (Figure 1). In the present case we chose to examine the performance of the anion exchanger DEAE (diethylaminoethyl) sepharose, used for the first of three chromatographic steps in the purification of Ciliary Neurotrophic Factor (rHCNTF) (Figure 1). DEAE resin is particularly useful for this first step, since many of the host cell contaminants

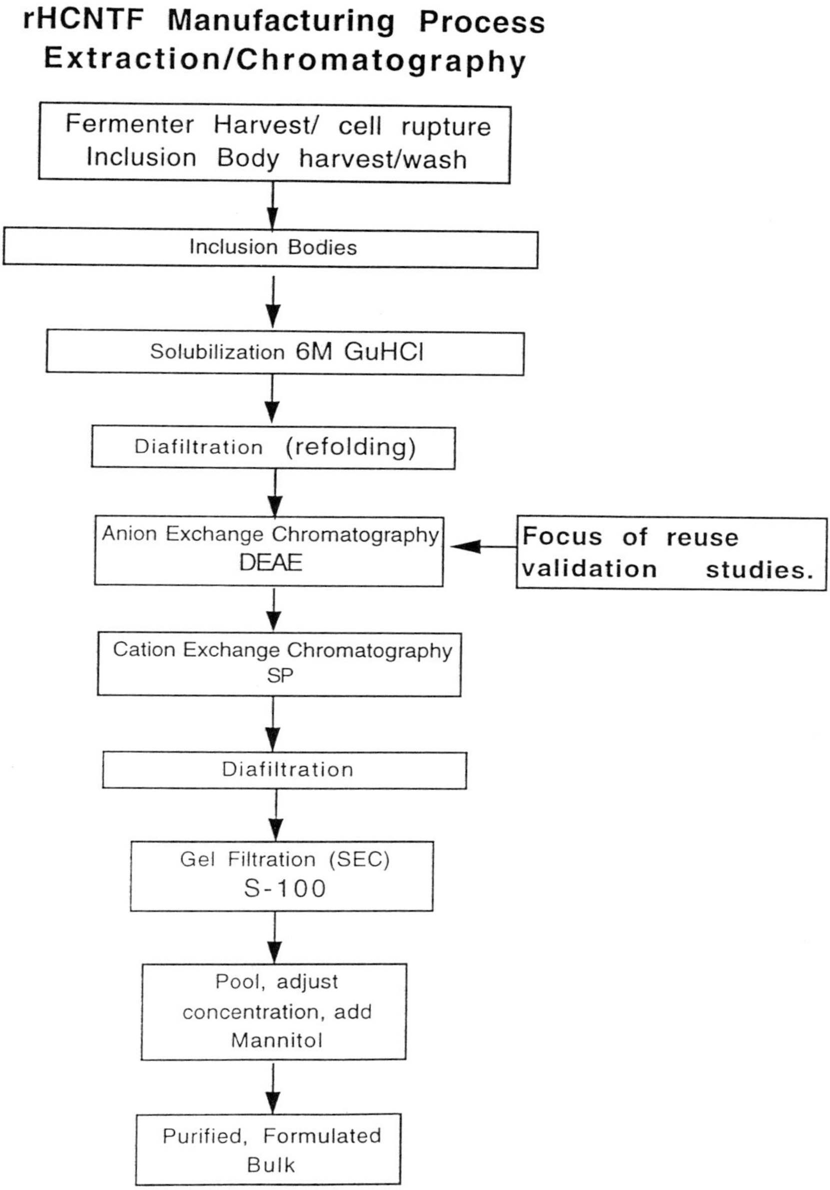


Figure 1 rHCNTF Manufacturing process outline.

such as lipoproteins, nucleic acids bind very strongly (*3,4,7*) to this matrix, whereas rHCNTF is eluted at a moderate salt concentration.

The scaled-down column was utilized to perform 30 chromatography cycles to satisfy a claim of 20 reuse cycles specified in the manufacturing operation. Following each purification cycle, the column cleaned and sanitized following the procedures utilized in the manufacturing plant. A number of performance parameters were examined to ensure consistent operation. The chromatograms were evaluated in terms of retention volume and peak volume to detect any deterioration in binding and elution characteristics. The step yield of the product was also monitored. The quality of the product produced was evaluated using a number of biochemical assays including SDS-PAGE, DNA assay, endotoxin assay and western blot analysis for *E. coli* protein contamination. The column flow-through during equilibration was routinely monitored for leachables generated from resin matrix using an anthrone-based carbohydrate assay (*8*). To expedite the analysis and conserve resources, we elected to analyze every fifth run and reserved analysis of flanking runs for discrepant results. At the conclusion of the study, a final blank run was performed, to ascertain whether there was any accumulation of contaminants on the resin.

Assay data was plotted in the form of run charts. The validation protocol required all of the data from each cycle to conform to manufacturing specifications. In addition, the data was analyzed statistically. Any outliers outside a range of $\pm$ 2 standard deviations around the mean needed to be investigated and explained.

## Materials and Methods

**Bench Scale Column System**. All small scale chromatography was performed at 4° C. A 14.1 x 1.0 cm column having an 11.1 mL volume Amicon (Beverly, MA) was packed with DEAE (diethylaminoethyl) sepharose fast flow resin from Pharmacia (Piscataway, NJ) to the same height as the manufacturing column. The system was driven by a Watson Marlow 101 U peristaltic pump (Watson Marlow Inc., Wilmington, MA). Absorbance at 280 nM was measured with a UV1 detector with 1.0 cm flowpath (Pharmacia LKB, Piscataway, NJ), and recorded with a REC 102 two pen chart recorder (Pharmacia LKB, Piscataway, NJ). All buffers were aliquots from large-scale batches which were subsequently used for actual manufacturing runs.

**Production Process.** Recombinant HCNTF (a 22.8 kD, monomeric protein) is expressed, following lactose induction, in RFJ26 pRPN40, an *E. coli* K12 bacterial host strain transformed with a plasmid containing the rHCNTF coding sequence, a Lac (UV5) promoter and Kanamycin ($Kan^r$) resistance. It exists in the form of non-native inclusion bodies in the host cells. The harvested cells are ruptured with a high pressure (550 bars) homogenizer (Niro Soavi, Parma, Italy). Inclusion bodies, released during cell disruption, are recovered using a continuous flow centrifuge Sharples AS 16VB (Alfa Laval Sharples, Warminster, PA), operating at 17,000g. The inclusion body paste is extracted using denaturants. The product is refolded by gradually exchanging the solution into 50 mM Tris, 0.05 mM EDTA, 1.0 mM DTT, pH 8.5 buffer using a diafiltration step with a 10,000 NMWCO kD cut off hollow fiber membranes (AG Technology, Needham, MA). rHCNTF emerges from these preliminary steps as a relatively pure (60-86% of total protein) protein in solution. Diafiltration removes residual chaotropes used for extraction, adjusts the pH and lowers the conductivity to allow resin binding. The refolded product exists in solution primarily as a monomer (>95%). It is microfiltered using a 0.22 u cellulose acetate membrane (Microgon, Laguna Hills, CA) and subjected to DEAE Sepharose chromatography for further purification.

As the first of three chromatographic steps, the DEAE column encounters the highest levels of many contaminants, including nucleic acids, residual host cell proteins and endotoxin. The proof of its resistance to repeated use and cleaning is paramount to any production reuse strategy (*1-4*).

**DEAE Chromatography Conditions.** A typical DEAE chromatographic production run (Figure 2), performed according to established manufacturing records, consists of the following steps:

**Initial Equilibration.** Removal of the column from storage in 0.01 N NaOH or caustic sanitization by column equilibration with 5.0 column volumes (c.v.s) of a high molarity equilibration buffer (500 mM Tris pH 8.5).

**Final Equilibration:** Accomplished with 5.0 c.v.s of a second low salt, low molarity buffer (50 mM Tris, pH 8.5) reduces the conductivity and prepares the column to receive the load.

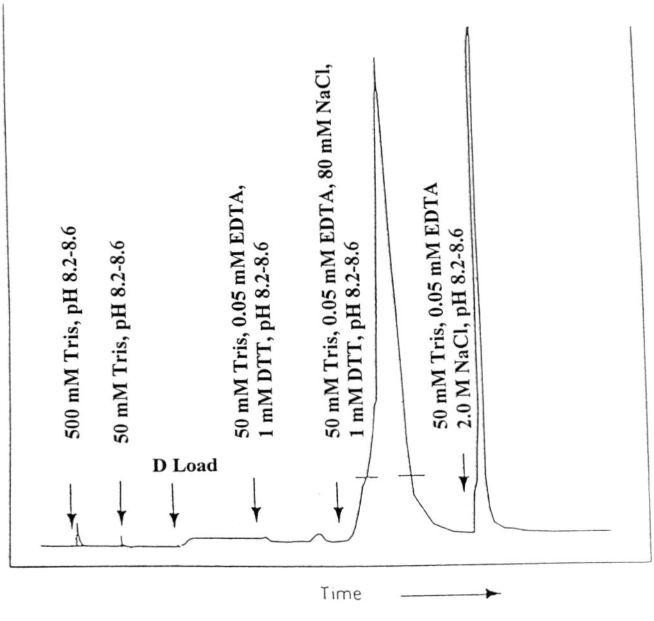


Figure 2 Manufacturing run DEAE elution profile detailing process steps.

**Loading.** The equilibrated column is loaded with refolded rHCNTF. The concentration of the load was approximately 0.5 mg/mL. The total amount loaded was 5-10 mg of protein per mL of resin. Load conductivity was < 3.0 mS/cm.

**Wash Step.** After loading, the column is washed with 5 c.v.s of wash buffer (50 mM Tris. 1.0 mM DTT, 0.05 mM EDTA pH 8.5) to remove unbound contaminants prior to elution.

**Elution.** Elution of rHCNTF from the resin is accomplished with a step gradient employing a moderate salt elution buffer (80 mM NaCl in 50 mM Tris, 1.0 mM DTT, 0.05 mM EDTA pH 8.5). Peak pools are collected when the observed absorbance at 280 nm exceeds 20% of full scale (2.0 AU 280 = full scale).

**Resin Regeneration/Sanitization.** Following elution, the column is stripped with high salt buffer (50 mM Tris, 2.0 M NaCl, pH 8.5), sanitized with cleaning solution (0.5 N NaOH) and stored in 0.01N NaOH prior to the next use.

Note: All equilibration, load, wash, retention peak pool, post peak and regeneration volumes were confirmed gravimetrically as per Manufacturing record instructions.

**DEAE Small-Scale Reuse Study.** The small-scale reuse study comprised a 900-fold scale down of the DEAE Sepharose Fast Flow production column (Table I) used as the first chromatographic step of rHCNTF purification. The column was scaled down linearly by reducing the diameter while maintaining the bed height. The column was loaded at 9.0-10.0 mg per mL of resin representing the upper end of the loading range employed in manufacturing. The flow rate was adjusted to reproduce the production linear flow rate and allow the appropriate resin contact (residence) time. Additionally, all process fluids, ion exchange resin and column starting materials were obtained from actual manufacturing runs to ensure that they were representative of the usual contact materials.

**Table I. DEAE Manufacturing Process Column Specifications**

| *Parameters* | *Specifications* |
|---|---|
| Retention Volume* | < 2.0 c.v.** |
| Peak Volume | <4.0 c.v. |
| DNA Clearance | $>1X10^3$ |
| Endotoxin Clearance | $>1X10^3$ |
| % Yield | $\geq$ 40% |

** Dimensions: 14.1 X 30 cm= 10.0 L c.v. (column volume).
* Volume of elution buffer passed before the peak collection begins.

The length of equilibration, column loading and post run regeneration/cleaning steps were all carefully defined so as to replicate those steps as they were performed in actual manufacturing. Manufacturing records detailing chromatographic operations were adapted for small-scale and used for chromatography and data recording. All buffers and protein solutions (DEAE column load and pool material) were stored frozen (-20°C) as single-use aliquots and were routinely subjected to extensive QC in-process testing, providing an added level of assurance as to their integrity. The column was operated by means of a peristaltic pump, with buffers, load material and cleaning solutions manually changed during chromatography. Likewise, all fractions and pools were collected manually based on manufacturing record instructions. Small scale reuse runs were performed over a five day period without column storage and columns were stored a minimum of 72 hours over each weekend during the study.

**Analytical Methods.** The following established assays were performed on the DEAE peak pool material.

**SDS PAGE.** Performed via the method of Laemli (*9*) using precast 15 % gels (Novex Inc. San Diego, CA). Gels were developed via silver stain with prepared reagents(Novex Inc. San Diego, CA).

**Protein Determination.** Protein was quantitated via the Bradford assay (*10*) with prepared reagents, (Biorad Inc., Hercules, CA).

**DNA Quantitation.** The amount of DNA in DEAE pool material was determined with the use of a Threshold analytical instrument (*11*), (Molecular Devices Corp., Menlo Park, CA).

**Endotoxin Quantitation.** Endotoxin remaining in the DEAE pool material was quantitated via the Limulus Amebocyte Lysate Assay, using prepared reagents (BioWhittaker Inc., Walkersville, MD).

## Results and Discussion

Figure 3 shows a comparison of a chromatogram from a manufacturing run with a chromatogram from a small scale run on an equivalent absorbance and volume scales. With regard to rHCNTF peak shape retention volume and peak volume, the correlation with the manufacturing run was good. In addition, the regeneration peaks for both runs are similar in both retention time and magnitude. Table II shows key performance parameters for the runs represented in figure 3, justifying scalability based on peak and yield performance parameters. Based on these results, we concluded that the scale-down column was operating with performance characteristics similar to the manufacturing scale. The small-scale system may then regarded as a reliable representation of the manufacturing process for the purposes of this reuse study.

Figure 4 shows elution profiles from three representative runs from the beginning, middle and end of the study. The chromatograms indicate that the load, wash, elution and regeneration components remained consistent throughout the course of the study. More detailed analysis of the performance parameters from every fifth run shows that retention volume, elution (peak) volume and percent yield all conform closely to their respective mean values (Figures 5a, 5b and 5c). This demonstrates retention of the resin's purification properties over many cycles with no deterioration due to caustic cleaning or contaminant accumulation.

The molecular integrity of rHCNTF and contaminant profile was analyzed with a number of standard assays including SDS PAGE (reduced and non-reduced), DNA analysis and endotoxin testing for selected reuse runs spanning the entire study. Silver-stained SDS PAGE gels, capable of detecting subnanogram levels of protein, are appropriate for demonstrating similarities and differences in DEAE

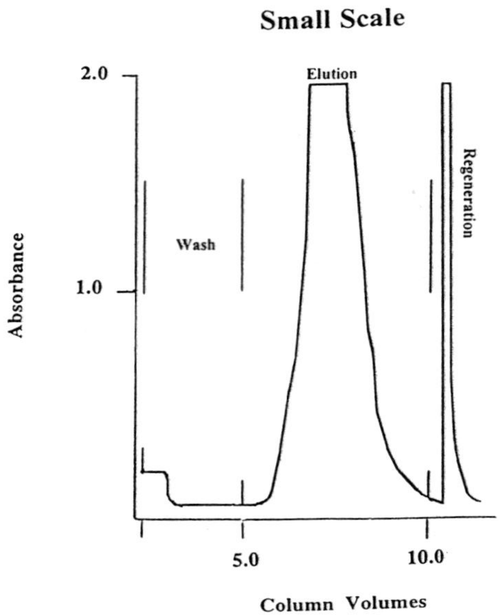


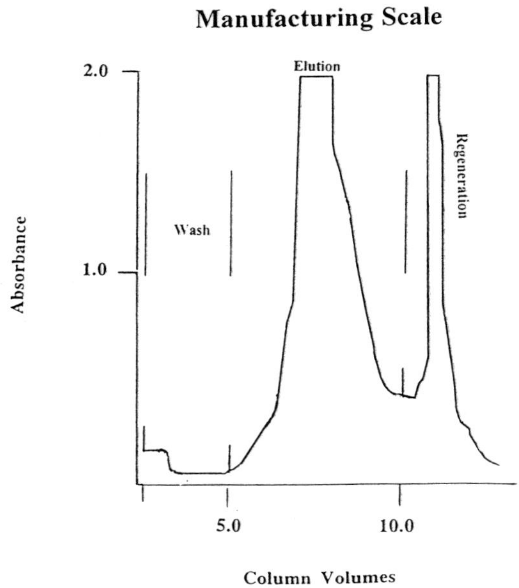


Figure 3 Comparison of small scale profile (upper chromatograph) and Manufacturing scale profile (lower chromatograph).

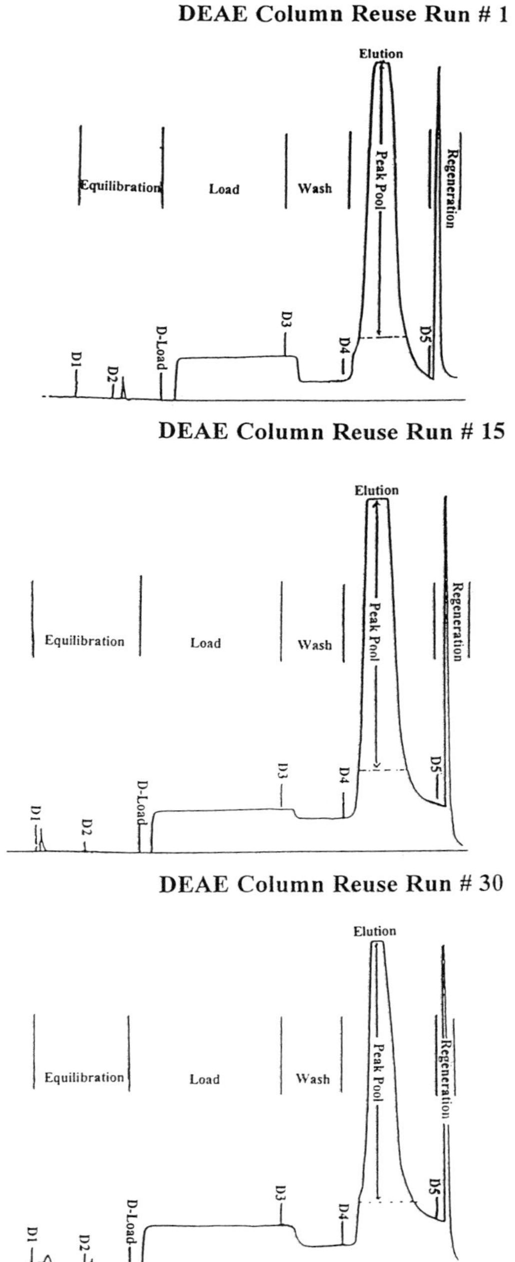


Figure 4 DEAE reuse run chromatographic profile comparison of 1, 15 and 30 demonstrating reproducibility throughout study.

5A

5B

5C

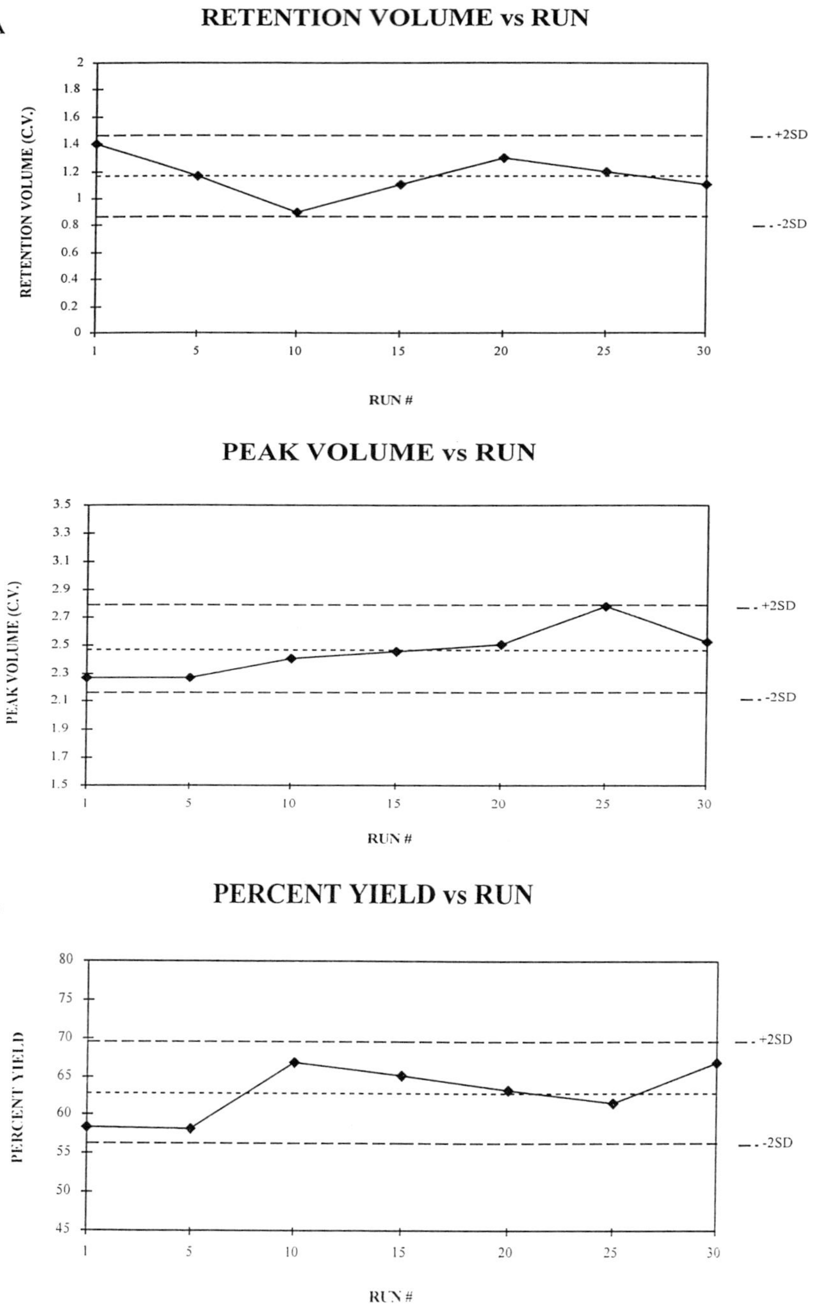


Figure 5 Comparison of retention volume (5a), peak volume (5b) and percent yield (5c) vs run over the course of the reuse study.

the first and last runs. The samples were prepared in reduced and non-reduced form by omission of dithiolthreitol from the latter sample. The gels reveal that rHCNTF was not degraded by purification with a resin having experienced 30 runs indicating no deleterious effects due to resin reuse. In non-reduced samples, there is evidence of a high molecular weight covalent dimer, arising as a result of disulfide bond formation between the single external cysteines of two rHCNTF molecules. There does not appear to be any appreciable increase in the proportion of this dimer to the monomer after purification with a resin used for 30 runs.

In general, the FDA requires the DNA levels in final parenteral products to be reduced to extremely low values (*3,6,7,*). DEAE resin is effective in removal of nucleic acids and endotoxin from bacterially derived rHCNTF, since both are highly negatively charged at the pH values used for anion exchange rHCNTF purification. Assays for these two benchmark impurities were included to assess this aspect of column performance during the study. DNA was quantitated via a Threshold assay (*4, 6*), while endotoxin was assayed by LAL gel clot method. DNA and endotoxin values for rHCNTF DEAE starting material are 2.8 x $10^6$ pg per mg and >8.6 x $10^5$ EU per mg rHCNTF respectively. DNA contamination in the DEAE pool was maintained below 41.8 pg/mg and endotoxin remained below .07 EU/mg (Table II). Clearance factors for each are maintained at consistently high levels (>$10^4$ for DNA and >$10^6$ for endotoxin) throughout the study (Table II). Both contaminants were consistently cleared to very low levels throughout the course of the study. As shown in the table, the actual DNA and endotoxin clearance achieved throughout the reuse study exceeded manufacturing specifications by at least a log.

**Table II. Comparison of DEAE Column Performance Parameters**

| Performance Parameter | Typical Manufacturing Run* | Range observed in the Reuse Study | Manufacturing Specifications |
|---|---|---|---|
| Percent Yield | 63% | 58.3-66.8% | >40 % |
| DNA Clearance | >$10^4$ | >$10^4$ | >$1x10^3$ |
| Endotoxin Clearance | >$1x10^6$ | >$1x10^6$ | >$1x10^3$ |

*These numbers are derived from an actual large scale manufacturing run.

Figure 7 shows the trend of absolute values of DNA and endotoxin levels of the DEAE pools obtained from every fifth run in the study. This data was analyzed statistically by identifying any data points outside the 2 standard deviation range. DNA contaminant levels in run #25 were found to be slightly

**Silver Stained SDS PAGE Gel From Reuse Run #1**

1) Low m.w. Stds. (reduced)
2) Blank
3) CNTF Std. (reduced)
4) D Load (reduced)
5) D Pool (reduced)
6) Blank
7) CNTF Std. (non reduced)
8) D Load (non reduced)
9) D Pool (non reduced)

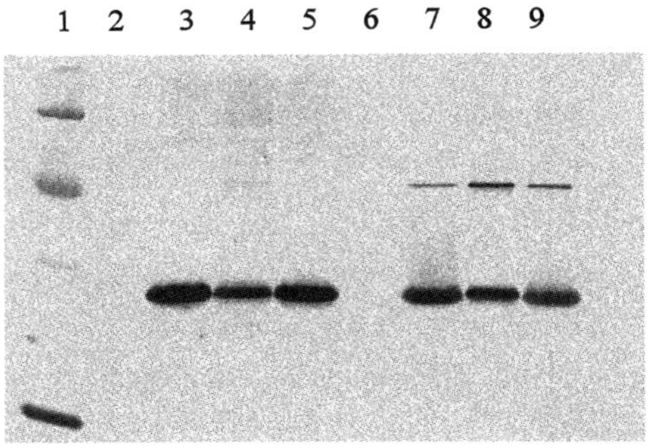


**Silver Stained SDS PAGE Gel From Reuse Run #30**

1) Low m.w. Stds. (reduced)
2) Blank
3) CNTF Std. (reduced)
4) D Load (reduced)
5) D Pool (reduced)
6) Blank
7) CNTF Std. (non reduced)
8) D Load (non reduced)
9) D Pool (non reduced)

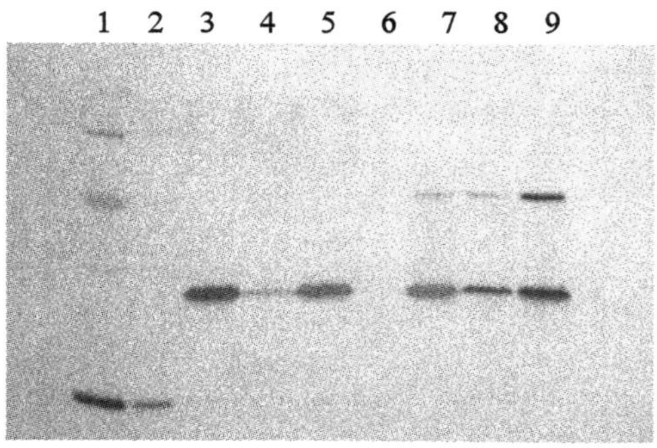


Figure 6 Comparison of the DEAE load and pool material for runs 1 (top) and 30 (bottom) demonstrating consistency of pool material.

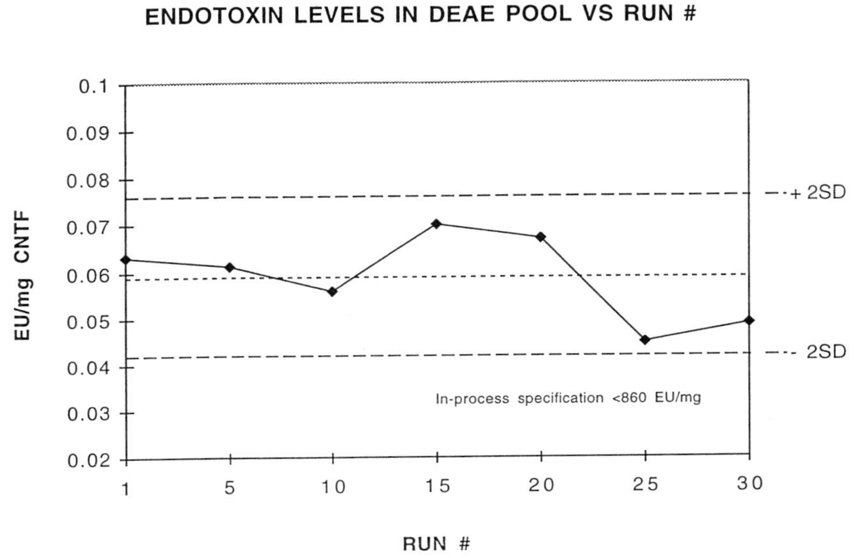


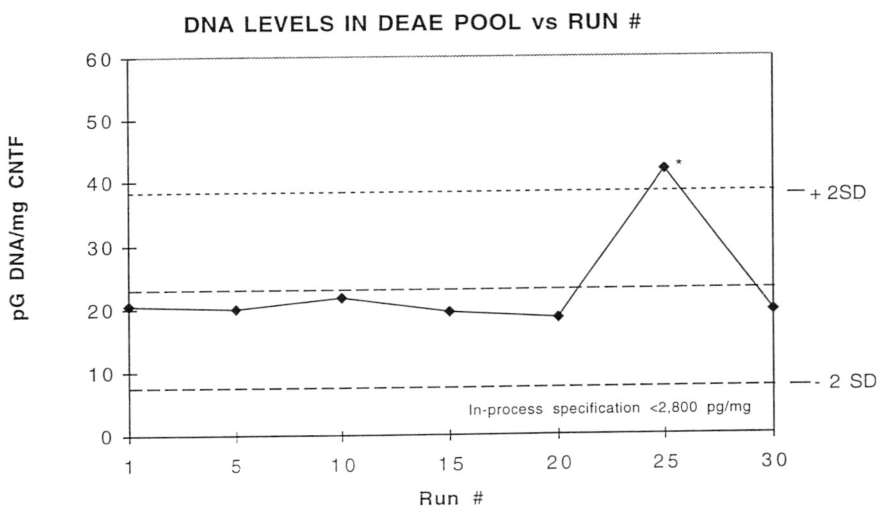


Figure 7 Analysis of DEAE contaminant levels throughout the reuse study for Endotoxin (top) and DNA (bottom) demonstrating retention of purification characteristics.

higher than the 2 S.D. range at 41.8 pg per mg protein. Although this level of clearance is still at least one log higher than the in-process specification of $10^3$, an investigation was carried out to identify the source of the variance. The investigation revealed an improper column wash step after loading as a probable cause. The analysis of run # 30 indicated DNA values of 19.3 pg/mg which is within the 2 S.D. range.

The release of resin leachables after NaOH exposure was investigated by anthrone analysis (*8*) of equilibration samples. The anthrone assay detects the presence of carbohydrates and can be employed to detect free carbohydrate moieties released as a result of sepharose degradation by NaOH cleaning and storage (*5,8*). Results for runs 1, 2, 15 and 30 were plotted versus column volume (Figure 8) during equilibration. These results demonstrate that, during equilibration prior to the first run, there is a high initial value of these column leachables which is not reduced below the level obtained in subsequent runs (2, 15 and 30) during equilibration. However, the anthrone reactive material during subsequent column equilibrations drops to <0.5 ug/mL levels. It is interesting to note that the run 1 values were obtained despite rigorous NaOH pretreatment, which should have removed residual material from fresh resin. This may have further implications for other Sepharose based resins, where this phenomenon may occur. As a result of these findings, a recommendation was made to the Manufacturing group to perform a blank run, with process buffers, on newly packed resin, prior to attempting any purifications.

To scrutinize contaminant or product accumulation during the study, a post study blank run was performed (Figure 9), substituting wash buffer for actual load material and collecting the column effluent during elution for analysis. As seen from SDS silver stain of the collected material (Figure 9), there was no detectable residual accumulation of any stainable material over 31 reuse runs. Silver staining is capable of detecting sub-nanogram quantities of protein. In fact, an inferred lower limit of 0.1 ng, would imply a level of < 10 ng/mL of any single protein band. This indicates that the resin was sufficiently cleaned after each reuse run.

## DEAE Blank Run Performed After Reuse Study

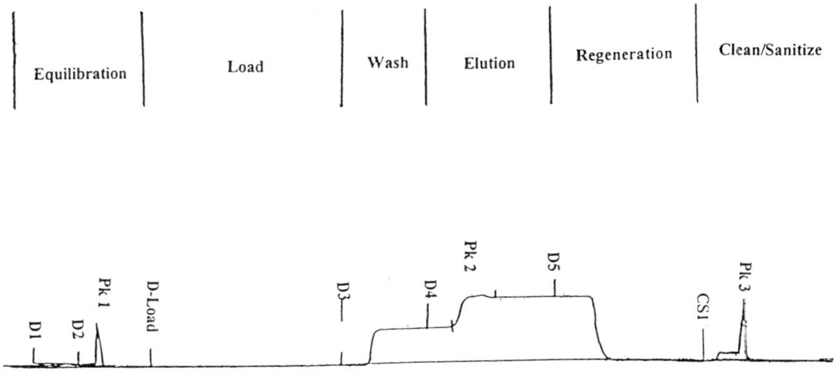


## Silver Stained SDS PAGE Gel From Blank Run (#31)

1) L.m.w. Stds.

2) CNTF Std.

3) Pk # 2

4) Pk # 1

5) Pk # 3

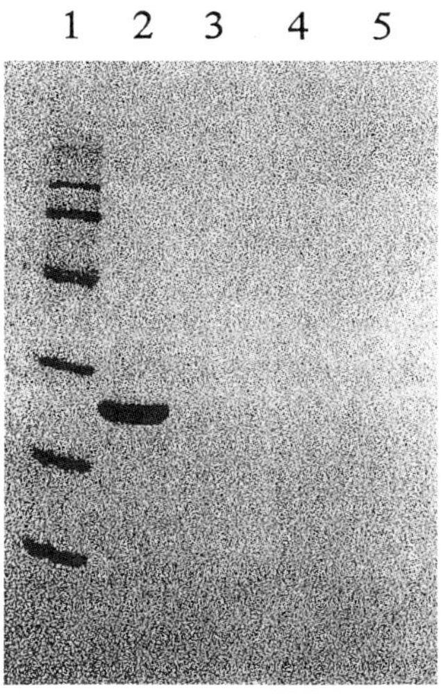


Figure 8 Illustration of blank run details: Blank run profile (top) followed by SDS PAGE analysis (silver stain) of collected peaks (bottom).

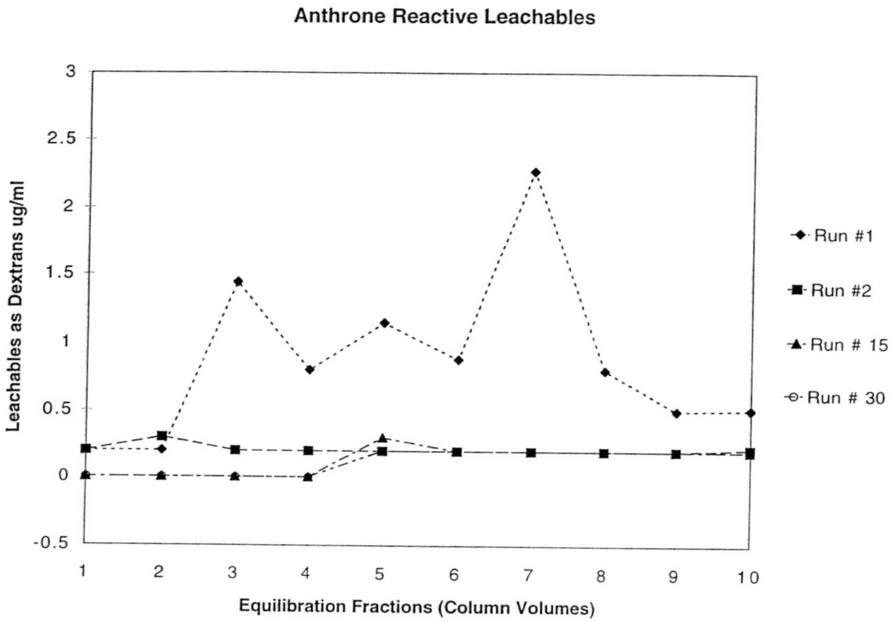


Figure 9 Analysis of release of anthrone reactive ( NaOH leachable) material as dextran over the course of the study (runs 1, 2, 15, 30).

## Conclusions

This reuse study clearly shows that the DEAE anion exchange column used for rHCNTF purification demonstrated reproducible, consistent chromatography over 30 small scale reuse runs. Physical column performance parameters such as peak retention volume, peak volume and yield are all faithful representations of the large scale chromatography and remained unaffected by repeated use of the resin. The rHCNTF molecular integrity appeared to be unaltered by any interactions with column resin throughout the study, indicating the DEAE chromatography media is resistant to chemical alterations from repeated exposure to caustic cleaning reagents employed in the manufacturing process. Clearance factors for endotoxin and host cell DNA remained consistent and were not altered by repeated column use. Final blank run data show no accumulation of contaminants or residual product during the study (as determined by silver stained SDS PAGE). This indicates that the resin was adequately cleaned between individual cycles. Finally, the resin was resistant to repeated NaOH sanitization as demonstrated by the absence of any increase in anthrone reactive material in column equilibration samples over the course of the study. Additional work, which augmented these results, involved contaminant challenge studies (for DNA and endotoxin clearance from spiked DEAE load samples) incorporated into post study experiments, in comparison with new resin of the same lot number. These studies demonstrated that the used resin retained its clearance capabilities for these test contaminants.

## Literature Cited


1. Adner, N.; Sofer, G. *BioPharm.* **1994**, *7(3)*, 44.
2. Seely, R.J.; Wight, H.D.; Fry, H.H.; Rudge, S.R.; Slaff, G.F. *BioPharm.* **1994**, *7(7)*, 41.
3. PDA Biotechnology Task Force on Purification and Scale-up. *J. Parental Science and Technology.* **1992**, *46(3)*, 87.
4. Dasarathy, Y. *BioPharm.* **1996**, *9(10)*, 41.
5. Johansson, B.L. *BioPharm.* **1992**, *5(3)*, 34.
6. Esber, E.C. "Points to Consider in the Purification and Testing of New Drugs and Biologicals Produced by Recombinant DNA Technology" Office of Biologics Research and Review, Center for Drugs and Biologics, 1985.
7. Petricciani, J.C.; Horaud, F. *Biologicals.* **1995**, *23*, 197.
8. Scott, T.; Melvin, E.H. *Anal. Biochem.* **1953**, *25*, 1656.
9. Laemli, U.K. *Nature (London).* **1970**, *227*, 680.
10. Bradford, M. *Anal. Biochem.* **1976**, *72*, 248.
11. Kung, V.T. et al., *Anal. Biochem.* **1990**, *187*, 220.

Chapter 11

# Start-Up and Validation of Sterile Formulation and Filling Processes for the Manufacture of Parenteral Aluminum Hydroxide-Based Vaccines

**R. A. Ramelmeier[1], P. M. McHugh[1,4], M. S. Rienstra[2], C. J. Orella[3], W. L. Stobart[1,4], M. W. Henley[2], and R. D. Sitrin[1]**

**[1]Merck Research Laboratories and [2]Merck Manufacturing Division, West Point, PA 19486**
**[3]Merck Research Laboratories, Rahway, NJ 07065–0900**

The start-up and validation of sterile processes for manufacturing parenteral vaccines involves a considerable effort across many disciplines. Using Merck's Hepatitis A vaccine (VAQTA) as a primary example, the start-up and validation challenges associated with sterile formulation and filling are described. Process robustness was generally determined employing a worst-case analysis of the critical parameters and their effect on the critical quality attributes, which were identified in prior characterization studies. The final validation of the processes was completed during production demonstration lots. The formulation process was validated based on the performance of the first ten consecutive lots, while the formal validation of the dilution and filling steps was based on 3 consecutive lots within the first ten.

While manufacturing parenteral formulations, the final sterile steps are particularly critical. Failure at this point results in a considerable loss of invested time and resources. Of utmost importance, however, is the quality of the final product, which is a human injectable. Consequently, the FDA has placed considerable emphasis on process validation for sterile products since the mid 1970's (*1,2*). Process validation provides higher assurance for batch-to-batch success and greater confidence that the product will meet its pre-determined specifications and quality attributes. When executed properly, validation can benefit a company by assuring a controlled process and a high quality product without significant testing in the long term.

According to the Pharmaceutical Manufacturer's Association (PMA), prospective validation (rather than retrospective) must be performed for the validation of sterile processes (*3*). Routine end-product testing is inadequate because it cannot assure product quality due to limited statistical sampling. This is particularly true for the sterility test where a product batch with a true 1.0% contamination would be

[4]Formerly in Bio/Sterile Validation in the Merck Manufacturing Division.

released eight times out of ten based on the USP sterility sample size of 20 units out of 1000 (*3*). Many aspects of the manufacturing process require close scrutiny via a well-documented validation program, including the equipment and control systems, the facility and its environmental conditions, raw materials and quality testing, personnel, and the process itself. This chapter will focus primarily on the sterile formulation and filling processes, their related equipment, and the unique start-up and validation challenges associated with aluminum-based vaccines. The other aspects of validation have been summarized thoroughly elsewhere (*4-6*).

## Background

The start-up and validation approaches discussed in this chapter are based on experiences with Merck's aluminum hydroxide-adsorbed vaccines for immunization against Hepatitis A (VAQTA) and Hepatitis B (RECOMBIVAX HB), with major emphasis on VAQTA (*7*). VAQTA is derived from an attenuated picornavirus that is highly purified, then inactivated in a low concentration of formaldehyde (*8*). RECOMBIVAX HB is a recombinant surface antigen from yeast (*9*). Both vaccines possess complex macromolecular structures.

The protein antigens are adsorbed to an aluminum-based adjuvant to improve immunological response; these adjuvants possess a long history of safety and efficacy in humans. (*10-12*). Currently, only the aluminum-based adjuvants have regulatory approval for routine injection into humans. The exact mechanisms for the improved immunological response are not known; a "depot effect" or enhanced recognition due to its large size are believed to play a major role (*13*). There exist a number of aluminum-based adjuvants (*11-14*). The adjuvant is created by precipitating aluminum into aluminum hydroxide lattices, which polymerize into higher-ordered structures over time. This polymerization releases hydrogen ions making long-term pH difficult to control (*11,12*). pH control is an important aspect of the validation effort as will be discussed below.

Aluminum-based vaccines present several challenges, making process development, characterization and scale-up particularly difficult. The first is the sterility requirement. Since the product cannot be sterile-filtered after adsorption to aluminum hydroxide, all process steps subsequent to the precipitation step must be carried out under aseptic conditions and preferably in a closed vessel. These formulations generally do not contain preservatives; sterility must rely completely on the sterile process.

To decrease risks during sterile processing, a closed system is the preferred route. To attain this added level of assurance, however, a considerable validation effort is required; the closures must be validated using pressure-hold, helium-leak, and/or microbial challenge tests, while the sterilization cycles must be validated using biological indicators. Closed systems can be sterilized and maintained in several ways. For fixed stainless-steel vessels and piping, the technology for SIP (sterilization-in-place) has been well-developed. Sampling and transfer from such systems require special valving configurations to steam-sterilize the connections. For small-scale bottles or stainless-steel cans (<20L), sterile-welder technology has provided a simple and flexible means for making connections without breaking a closed system; this is particularly effective for making clinical batches where more flexibility is often needed. The technology uses special tubing (C-Flex; 1/8 in. ID) and a qualified heat-welding device for making closed-system connections. The disadvantage of this technology is the tubing size limitation; however, the technology for tubing sizes up to 1/2 and 1" ID should be available in the near future.

The most significant challenge arises from the complex nature of aluminum-based adjuvants and protein antigens and their interactions with each other. Although considerable research has been carried out to characterize aluminum-based adjuvants and aluminum hydroxide-adsorbed proteins and vaccines, these products remain poorly defined (*11,12*). Due to the complexity of the vaccines and these adjuvants, much of the work is empirical. Table I lists some of the current analyses available for characterizing aluminum hydroxide properties. The values from these analyses can be used to check process consistency and robustness; unfortunately, one cannot use the properties to effectively optimize such systems as little data exists which correlate these properties with product efficacy.

## The Processes and Equipment

**Formulation via aluminum hydroxide coprecipitation.** Figure 1 shows a process for making aluminum-based vaccines from a final purified bulk through filling. The first sterile step is formulation, which for aluminum-based vaccines can be via aluminum hydroxide coprecipitation. The co-precipitation process begins with the addition of a low pH aluminum salt solution to the purified protein antigen in buffered saline. A caustic solution is then added to precipitate the aluminum. As the pH rises and the aluminum hydroxide precipitates, the protein antigen is adsorbed. Base is added until the aluminum hydroxide suspension reaches a suitable pH, such that the final product pH stabilizes within physiological conditions (near 6.0). Excess salts and other soluble components remain in the supernatant following the co-precipitation. Since the aluminum adjuvant is sufficiently more dense than water, the aluminum hydroxide-adsorbed vaccine can be "washed" of the residuals through a series of settle/decants and re-suspensions with physiological saline, or alternatively via diafiltration. Washing is carried out until the residuals are sufficiently cleared.

The coprecipitation vessel is jacketed to maintain the vaccine at cool temperatures (2-8 °C). The vessel is equipped with ports and dip tubes for adding reagents, sampling, transferring product, and decanting supernatant above the settled aluminum hydroxide-adsorbed vaccine. Agitation is accomplished using an impeller mixer, a magnetic stir bar, or a magnetically-driven turbine impeller depending on the scale of operation. The coprecipitation to make VAQTA was carried out in a sterile multi-port glass container with C-Flex connections . A sterile welder device was used to maintain a closed system. To attain a consistent final pH, an automated pH titration system was developed (see Figure 2).

**Dilution.** Dilution is a fairly straightforward process step. Starting with a concentrated bulk, the amount of "diluent" needed is calculated and combined with the bulk to achieve the desired final product concentration. The diluent is often manufactured to have a buffer composition identical to the concentrated bulk, so that the only concentration changed upon dilution is that of the active ingredient. After mixing, the product is sampled and assayed to confirm that the desired final concentration has been achieved.

**Table I. Analyses Available for Characterization of Aluminum Hydroxide Products**

| Property | Analysis |
|---|---|
| **1. Particle Size** | |
| - Primary | Rate of acid neutralization |
| - Secondary aggregates (1-10 microns) | Light scattering |
| - Settling aggregates | Settling rates |
| **2. Morphology/Structure** | IR, NMR and other spectroscopic techniques<br>Electron Microscopy |
| **3. Point of Zero Charge (PZC)** | Zeta potential as a function of pH |
| **4. Protein Adsorption** | Protein assay, bio-activity assay |
| **5. Ion/Chemical Composition**<br>- Aluminum<br>- Phosphate<br>- Sulfate<br>- Borate | ICP, colormetric titration, ion chromatography |

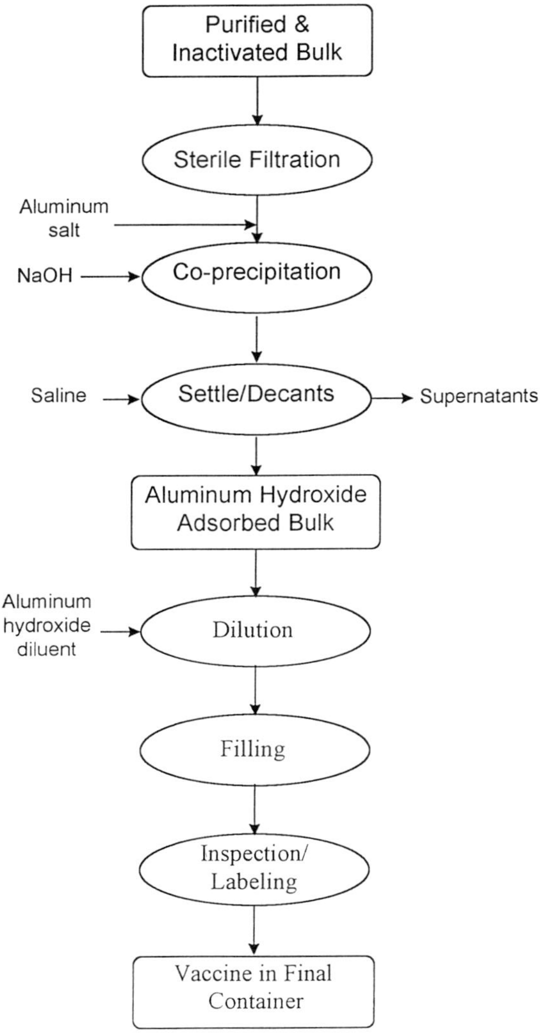


Figure 1. Formulation and Filling of Aluminum-based Vaccines.

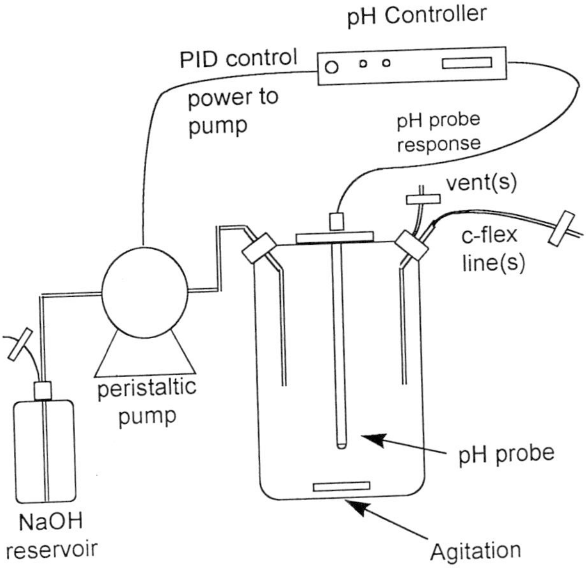


Figure 2. General Schematic of the Aluminum Co-precipitation Vessel for VAQTA.

The equipment for a dilution system at manufacturing-scale typically consists of the bulk vessels containing the concentrated active ingredient and diluent, a dilution vessel with mixing capability, a scale or load cell, silicone tubing or stainless steel piping, and a pump or pressure delivery system for transfer. The dilution vessel should be chosen to accommodate the range of possible diluted batch sizes. Also, an estimate of possible deviation from the target final concentration should be made, based on scale accuracy, accuracy of the charge method, losses during the charge, and assay variability.

For high levels of dilution, accuracy can be improved by using serial dilutions. For example, to reach a 1:1000 final dilution level, three consecutive dilutions of 1:10 may be made; assays are generally employed to confirm the proper dilution at each step. Density should also be considered if charging by weight, since target final concentrations are normally in terms of volume. Also, mixing in the dilution vessel should be considered, especially if the dilution vessel is to also serve as the bulk product vessel for filling. Finally, the system must be sterilizable.

For VAQTA, a 100 L mixing vessel was chosen as the dilution vessel. The target final concentration could be reached with sufficient accuracy using a single dilution (i.e. serial dilutions not needed). The aluminum hydroxide diluent and concentrated vaccine bulk contained identical levels of aluminum. First, the desired diluted batch size was chosen, the required amounts of formulated vaccine and diluent were calculated, and the 100 L vessel was set-up and checked on a calibrated floor scale. Aseptic tubing connections were then made, and the required amount of formulated vaccine was then charged to the dilution vessel. Lastly, the diluent was charged to the dilution vessel and the final diluted bulk was mixed prior to sampling. All of these steps were conducted under a class 100 laminar-flow canopy.

**Filling.** Filling is the dispensing of homogeneous final diluted product into vials or syringes in accurate volumes, to ensure that the appropriate dose is consistently delivered. This is achieved by transferring product from a bulk holding vessel through the filling machine. The machine consists of pumps that are calibrated to deliver accurate volumes through dispensing needles.

**Mixing:** Mixing can be challenging for aluminum-based vaccines and cannot be overlooked. The filling process begins with the mixing of the bulk product vessel (often the same vessel as the dilution vessel). One cannot expect the filling system to deliver a well-mixed and consistent product if the product is not homogeneous before entering the filling system. A mixing protocol must therefore be developed such that the product can be made homogeneous in the bulk product vessel prior to filling, and can be maintained homogeneous throughout the filling operation (i.e. while the vessel volume vessel is continuously decreasing). Storage time and volume should also be considered for suspension products, as the settled component may become compacted in the bottom of a vessel over time.

For VAQTA, a design of the 100 L bulk product vessel was selected in which mixing was achieved solely through recirculation of the bulk through dip-tubes. This design avoided the challenge of maintaining sterility in vessels with shafts or mechanical seals and ensured proper mixing until the vessel was completely empty. A schematic of this vessel and the filling system is shown in Figure 3.

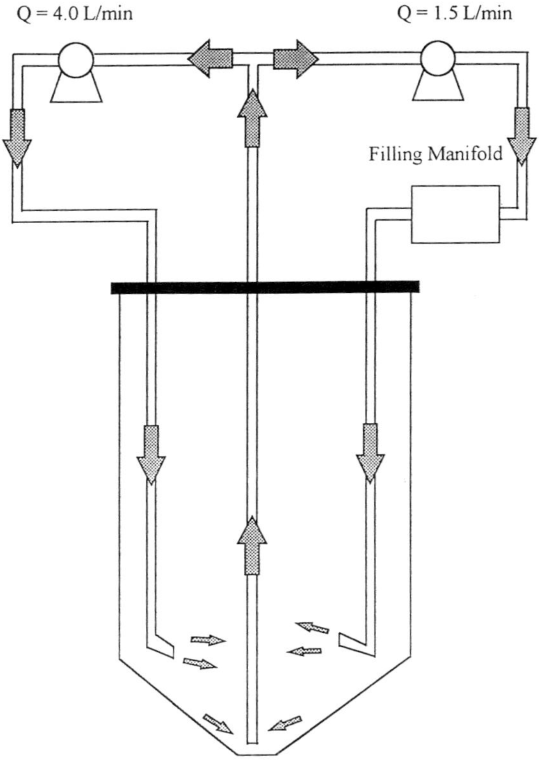


Figure 3. Recommended Filling Parameters for VAQTA.

**The Filling System**: Once a homogeneous state is established in the bulk vessel, the filling system must accurately and aseptically deliver a set volume to the vials or syringes. Delivery of homogeneous product is often aided by recirculation loops on the bulk vessel, with the filling system drawing product from a recirculation loop. The filling system should also provide minimal exposure of open sterilized vials or syringes to the environment. A variety of filling systems exist; the appropriate system for a given application depends on the final container type (vials or syringes), as well as the requirements for speed, accuracy, and level of automation. There is typically a trade-off between accuracy-of-fill, volume-of-fill, and speed. Filling pump types include diaphragm, peristaltic, and piston; time-pressure and mass-flow methods can also be used. The pumps are normally adjustable and accurate within a specified range of fill volumes. Other considerations when choosing a filling system include the level of maintenance required, cleaning, sterilization, system hold-up volume, and setup losses.

As seen in Figure 3 for VAQTA, a second recirculation loop was used to divert product through the filling manifold and back to the mixing vessel. This additional loop assured that homogeneous product was delivered to the filling machine. Both loops were controlled through peristaltic pumps. The filling manifold (TL Systems, Inc.) contained six filling pumps that drew product from the second recirculation loop and dispensed it into vials with a target fill volume of 0.70 mL/vial. Filling was carried out under a class 100 laminar-flow canopy.

## Start-up and Validation Approach

The validation of VAQTA involved a prospective approach, which can be broken down into 3 primary steps. For new products like VAQTA, validation begins in research and development (R&D) with the initial product and process characterization and optimization. This includes definition of the necessary unit operations, raw material needs, analytical methods, and process performance limits. R&D also identifies the critical product quality attributes and process parameters. The second step is the process validation study which involves further robustness testing with respect to the critical parameters; this defines the operational limits which do not adversely affect the critical quality attributes. Finally, the validation process is completed by running a minimum of 3 consecutive demonstration lots at full-production scale.

As described in Chapter 1, start-up and validation require a team effort. At Merck, members of various groups, including research and development, engineering/technology support, production, validation, quality control, regulatory affairs, and quality assurance, all participate in start-up and validation activities. For VAQTA, the characterization and pre-demonstration validation studies were performed by the R&D group, while manufacturing-consistency lots were carried out by production staff under the supervision of R&D, manufacturing support, and validation. For the critical validation activities, such as aseptic dilution and filling, the validation group designed and authored the validation protocols.

Table II lists the important variables which control and measure performance for the formulation, dilution and filling of aluminum hydroxide-adsorbed vaccines, with emphasis on VAQTA. The Critical Process Parameters are the important process variables which affect the process outcome. The Critical Quality Attributes provide the corresponding critical measures of process outcome or performance.

## Start-up and Validation of the Formulation Step

Of all the sterile steps, formulation is the most complex. It includes several product and buffer transfers, mixing, and a chemical reaction to form the aluminum-based adjuvant. Prior to start-up of VAQTA, a major effort was carried out to better understand aluminum precipitation and the automated pH titration process.

**The Characterization phase.** As outlined in Table II, three key parameters were considered for "validation" of the coprecipitation process . These variables were 1) mixing rate, 2) base addition rate, and 3) proportional band width (PBW) for automated titration. The base addition rate and agitation affect the time necessary to fully disperse base in the vessel, which must not exceed the controller response time. The band width is the pH interval prior to the pH set-point at which the control switches base addition from full "on" to PID control. For example, for a bandwidth of 2 and a set-point of 7.3, the controller switches to PID control at pH 5.3. Other significant variables, such as temperature, target titration pH, buffer concentration, and post-precipitation incubation time, were either fixed or optimized early in development to obtain desired product attributes or to be consistent with other coprecipitation processes at Merck. In addition, parameters related to scale-up and equipment design (such as location of NaOH dip tube, subsurface versus above surface discharge, mixing by magnetic stir bar versus impeller, etc..) were also considered, but they were generally selected based on prior experience and then readjusted as needed.

The primary goal of the early characterization work was to determine the conditions which would provide a stable pH. Several studies were executed in small-scale mixing vessels to determine the key parameters that affect aluminum hydroxide final pH. Results indicated that a reasonably high pH titration target (6.8-7.2, but not higher than pH 8.0) and addition of buffers were required to overcome the pH drift that occurs after the aluminum-based adjuvant is first formed. Mixing rate and NaOH addition rate had the greatest effect on pH drift. Due to the difficulty of scaling-up mixing operations, these variables had to be re-tested in the actual process vessel. Temperature also had a significant effect on pH drift and pH measurement of solutions containing aluminum-based adjuvant. Although the co-precipitation reaction is carried out isothermally (2-8°C for best vaccine stability), temperature can be particularly important if measuring the pH of a newly precipitated product off-line.

Other key variables which affect the final pH are the volume and concentration of NaOH and aluminum salt. The aluminum target is fixed to meet the desired aluminum concentration in the final product. The NaOH target is linked to the final pH, therefore it was not considered a critical parameter; the amount of NaOH added, however, was monitored for in-process consistency. As much as 5% variation in the solution concentrations is permitted based on their release specification. Because pH measures the progress of the reaction, these small variations will not affect the process.

**The Pre-Demonstration Validation Studies:** Operational limits were defined from practice runs employing the prototype of the manufacturing process vessel and automation equipment (Valley Instruments 506M pH controller with Ingold pH probe). A general schematic of the system is shown in Figure 2. The tests were carried out in the absence of the protein antigen. Since the antigen concentration in the bulk formulation was very low (<2.0 mcg/mL), these runs should effectively represent the final process. Overall, the tests proved extremely valuable. They provided information to accurately tune the pH controller, challenged the robustness of the control system and the co-precipitation process, and further characterized the manufacturing-scale process. The aluminum suspensions generated in these test runs were extensively characterized (point-of-zero-charge, particle size, settling rates and other properties). These analyses were compared to those of the formulation bulk for VAQTA and other Merck aluminum hydroxide products to bridge the practice and demonstration runs.

**pH**. Figure 4 shows a pH profile for a typical aluminum hydroxide precipitation test run. The procedure and component concentrations were identical to those used in the final manufacturing process. Table III lists the results of all of the test runs. The objective of the tests was to determine process robustness with respect to worst-case conditions of the key parameters: low agitation/poor mixing, fast NaOH-addition rate, and narrow band-width. A high NaOH addition rate, 90% of the pump maximum, with low agitation (225 rpm) and a PBW of 1, lead to a steady approach to set-point pH without considerable pH overshoot. The only failure during the practice runs resulted when agitation rate at these conditions was lowered to 100 rpm. With adequate agitation the process was extremely robust and a final precipitation pH of $7.2 \pm 0.1$ could be attained.

The final operational limits for the critical parameters were chosen to provide a rapid ascent to the set point pH (less than 30 minutes) without significant overshoot. The conditions are summarized as follows:

1. Proportional band width of 2.0
2. Agitation with free stirrer at 300-400 rpm, with 350 rpm target.
3. NaOH addition rate of 40-60% of pump maximum, with a target of 50% with a free magnetic stirrer.

These conditions reflect modifications made as a result of a change from a fixed spinner to a free stir bar during the first demonstration runs. This change resulted in only slightly poorer mixing and had a minimal impact on the overall process. The NaOH addition rate was decreased from the original target of 60% to 50% to account for the poorer mixing. The spinner was abandoned because its sterility could not be validated.

Other considerations relating to pH required special attention during the start-up. These included manual NaOH-addition, offset from the set-point pH, and integrity of the pH probe. A method for manual NaOH addition was developed as a back-up to the automated titrator; its performance was confirmed under standard conditions (Table III). A titration curve was generated which correlated pH with the amount of NaOH added, which can be used as a guide during the manual procedure.

In all of the test precipitations, the final steady-state pH fell short of the set-point pH. This was attributed to a slow NaOH-pump response to the controller signal.

**Table II. The Critical Parameters and Quality Attributes for the Formulation and Filling of Aluminum-Based Vaccines**

| Key Steps | Critical Quality Attributes | Critical Process Parameters |
|---|---|---|
| Formulation via aluminum hydroxide co-precipitation [a] | • pH | • Base addition rate<br>• Control band width (for automated pH control)<br>• Mixing rate |
| | • Sterility | |
| | • Alum conc. | |
| Dilution | • Aluminum/Antigen conc.<br>• Homogeneity | • Mixing rate<br>• Mixing time<br>• Volume ratio<br>• Equipment accuracy |
| | • Sterility | |
| Filling | • Aluminum/Antigen conc.<br>• Homogeneity | • Mix/recirc. Rate<br>• Manifold recirc. rate<br>• Filler speed/settings<br>• Alum/bulk age |
| | • Fill Volume | • Air entrainment<br>• Filler settings |
| | • Sterility | |

a) Antigen concentrations and alum properties such as point-of-zero charge (PZC) are important variables to measure consistency, but their absolute values are not critical to the formulation process.

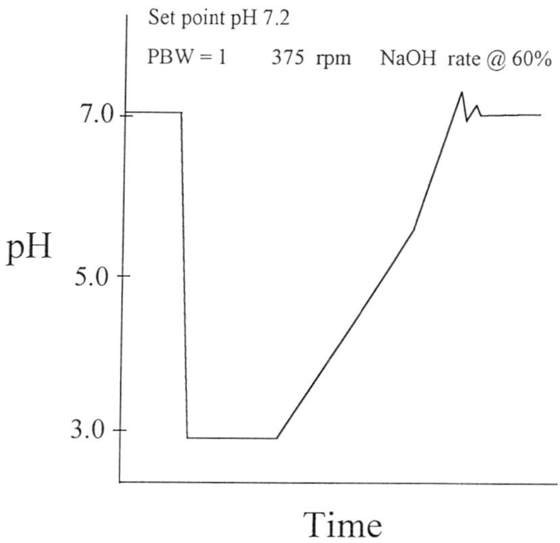


Figure 4. Aluminum Hydroxide Precipitation Test Run.

**Table III. Summary of Test Aluminum Hydroxide Precipitations**

| NaOH Flow Rate % of pump maximum | Agitation rpm | PBW | Set point pH | Steady State pH | Time for Steady State (min.) | Max. Transient pH |
|---|---|---|---|---|---|---|
| 30 | 300 | 1 | 7.0 | 6.80 | 25 | 6.8 |
| 35 | 375 | 1 | 7.2 | 7.10 | 30 | 7.1 |
| | | | | | | |
| 60 | 100 | 1 | 7.2 | 7.17 | 20 | 8.1 |
| 60 | 225 | 1 | 7.2 | 7.10 | 10 | 7.5 |
| 60 | 375 | 1 | 7.2 | 7.15 | 15 | 7.3 |
| 60 | 375 | 1 | 7.3 | 7.25 | 10 | 7.3 |
| 65 | 375 | 1 | 7.0 | 6.90 | 10 | 7.1 |
| 65 | 375 | 1 | 7.2 | 7.13 | 10 | 7.1 |
| | | | | | | |
| 90 | 100 | 1 | 7.2 | 9.2 | overshot | |
| 90 | 225 | 1 | 7.2 | 7.11 | 10 | 7.7 |
| 90 | 375 | 1 | 7.2 | 7.11 | 10 | 7.2 |
| 95 | 375 | 1 | 7.2 | 7.10 | 10 | 8.0 |
| | | | | | | |
| 60 | 375 | 2 | 7.3 | 7.21 | 10 | 7.21 |
| 60 | 375 | 2 | 7.3 | 7.20 | 10 | 7.2 |
| 60 | 375 | 2 | 7.3 | 7.20 | 10 | 7.2 |
| | | | | | | |
| 35 | 350 [a] | 2 | 7.3 | 7.20 | 20 | 7.2 |
| 50 | 350 [a] | 2 | 7.3 | 7.20 | 10 | 7.2 |
| 60 | 300 [a] | 2 | 7.3 | 7.20 | 10 | 7.2 |
| 60 | 350 [a] | 2 | 7.3 | 7.20 | 10 | 7.2 |
| 65 | 350 [a] | 2 | 7.3 | 7.20 | 10 | 7.2 |
| | | | | | | |
| 10 [b] | 350 [a] | n/a | 7.3 | 7.20 | 25 | 7.2 |
| 30 [b] | 350 [a] | n/a | 7.3 | 7.30 | 15 | 7.3 |
| 60 [b] | 350 [a] | n/a | 7.3 | 7.20 | 10 | 7.2 |

a) Free magnetic stir bar
b) NaOH added manually

As the set-point pH is approached, the power-on interval to the pump becomes progressively shorter. Within 0.1-0.2 pH units this interval becomes so short that the pump is not allowed enough time to advance and pump NaOH. Adjusting the offset potentiometer in the controller failed to solve the problem. Since the offset was consistently 0.1 pH units below the set-point pH, the final set point was adjusted from pH 7.2 to 7.3.

Since pH measurement is critical to the process, reliability of the on-line pH probe must be demonstrated prior to each run. A series of pH probe checks and calibrations were established to identify a probe failure. If the failure is discovered prior to charging product into the vessel, a back-up vessel and probe is used. After product is charged to the vessel (pH 7) and aluminum salt is added (pH 3), the probe performance is verified. If the pH is within 0.5 units of target after these transfers, the automated process is used. If the pH probe does not respond acceptably, the manual process is employed. During the first 10 manufacturing lots, the pH titration system performed reliably; all faulty probes were identified prior to transfer of the bulk product into the vessel.

**Aluminum and Antigen concentration**: While not as critical as pH, the other quality attributes listed in Table II required consideration. The amount of aluminum-based adjuvant and thus the aluminum content in the bulk product were fixed. The aluminum limits for VAQTA were set based on pre-determined limits for aluminum hydroxide products at Merck. The antigen concentration was measured, but not targeted to a pre-determined level at this step. The final antigen content was adjusted during the subsequent dilution step.

**Sterility:** The glass vessel with C-Flex connections and pH probe was sterilized by a validated autoclave cycle. To provide sterility assurance in the closed vessel, both helium leak and microbial submersion testing were successfully performed. Since these tests proved that the vessel interior could not be contaminated by sources on the vessel exterior, additional challenges, including media challenges, were not conducted. Because the co-precipitation process occurs entirely in a closed system, the step is carried out in a class 10,000 environment.

**The Demonstration phase**: To complete the validation program, the process was "demonstrated" in the actual manufacturing environment. The final validation targeted the critical quality attributes of pH, aluminum concentration and sterility. Other attributes, such as antigen yield and aluminum hydroxide properties including PZC, provided positive indicators for consistency. Data from ten manufacturing lots were evaluated to demonstrate consistency in the formulation step. The first 5 lots were required to fine-tune the scaled-up process and the new equipment. The validation consistency lots for the final filling process were 6, 7 and 8. After final approval of the process and sign-off of the validation report, these lots were tested in the clinic.

Table IV shows the results of the co-precipitation process for 10 manufacturing lots. The post-precipitation pH (7.2 ± 0.1) and the total NaOH added were consistent; the amount of NaOH added was statistically identical to that observed during the simulation runs (212 ± 4 g, based on 8 runs). One demonstration lot resulted in significant pH overshoot, requiring pH adjustment (lot 3); this was due to faulty agitation unrelated to the automated pH titration. Safeguards were established (e.g. NaOH reservoir with pre-weighed amount of NaOH and a vortex check for mixing) to assure success in subsequent runs. Table V shows the off-line pH measurements of the final precipitated product, which are well within the aluminum-adjuvant pH specification of 5.5-7.0.

The aluminum levels in the manufacturing lots and the test runs are shown in Table VI. The aluminum levels show minimal statistical variation and are well within historical limits (0.35 to 0.62 mg/mL). Process yields of the combined inactivation-formulation steps were consistently high (105 ± 21 %); the variation coincides with the variance associated with the EIA assay. Finally, sterility testing of all the bulk aluminum hydroxide products was satisfactory.

To further verify that the practice runs were representative of the final process for VAQTA, the concentration of constituent ions (data not shown), PZC, and the particle size of the aluminum-based adjuvants were measured and compared for randomly selected lots (see Table VII). These properties statistically demonstrated minimal variation lot-to-lot and between the various aluminum hydroxide sources, including aluminum hydroxide diluent which utilizes the same precipitation process, but at a larger scale (300 L tank).

As discussed earlier, a settle-decant process is employed to clear residual salts and formaldehyde. The objective of this step is to achieve 100-fold clearance of residuals. This was optimized by properly balancing the decant volume against the number of decants This balance depends on the settling rates of the aluminum adjuvant. A large decant volume would require considerably longer settling times and result in the adjuvant compacting at the vessel bottom. A small decant volume would require more manipulations and buffer transfers. The best compromise for VAQTA was 6 supernatant decants comprising 54% of the total volume. The decant-line length was sized to allow 54% of the process volume to be withdrawn without disturbing the settled product. Table VIII shows the reduction in formaldehyde during the decant process and the variation from batch to batch. The final formaldehyde concentrations approached the 100-fold clearance target, well below the allowable levels. Some variation was expected due to minor variability in the vessel dimensions.

## Start-up and Validation of the Dilution & Filling Steps

**Characterization and Pre-Demonstration Validation Studies.** Once the 100 L mixing vessel was selected, initial studies were performed with RECOMBIVAX HB, whose particle size and settling characteristics are very similar to those of VAQTA . As seen in Figure 5, it was determined that a 90L bulk that had been allowed to settle for 14 days could be satisfactorily re-suspended after 15 minutes of recirculation at a rate of 5.6 L/min.

Based on these initial findings, simulated fills were carried out using a 30 minute re-suspension time and a minimum total recirculation rate of 5.5 L/min. The objective of these fills was to determine process robustness and the upper limit of the recirculation rate. These developmental fills used water, aluminum hydroxide diluent

**Table IV. Process Parameters and Precipitation pH for the Demonstration Lots a)**

| Lot | Base Flow rate (% of pump max.) | Agitation rpm | Base Added[c] g | On-line pH at 8°C | | | |
|---|---|---|---|---|---|---|---|
| | | | | pH of FIB[d] | pH after alum salt added | Max. pH (transient) | Steady-state pH |
| 1 | 60 | 350 | 204 | 7.17 | 3.17 | -- | 7.17 |
| 2 | 60 | 350 | 210 | 7.15 | 3.19 | 7.8 | 7.19 |
| 3 | 60 | 350 | 243 + 25g 1 N HCL | 7.20 | 2.99 | 8.7[e] | 6.60[e] |
| 4 | 60 | 350 | 223 | 7.26 | 3.04 | 7.2 | 7.20 |
| 5 | 60 | 350[b] | 211 | 7.26 | 2.92 | 7.4 | 7.17 |
| 6 | 60 | 350[b] | ~220 | 7.22 | 2.83 | 7.3 | 7.31 |
| 7 | 60 | 350[b] | ~220 | 7.10 | 2.80 | 7.4 | 7.19 |
| 8 | 60 | 350[b] | 218 | 7.28 | 2.93 | 7.5 | 7.18 |
| 9 | 40 | 350[b] | 220 | 7.25 | 2.88 | 7.2 | 7.18 |
| 10 | 50 | 350[b] | 211 | 7.27 | 3.02 | 7.3 | 7.20 |
| **Avg.** | **n/a** | **n/a** | **216 ± 6** | **7.2 ± 0.1** | **3.0 ± 0.1** | **7.4 ± 0.2** | **7.2 ± 0.1** |

**a)** Proportional band width is 2; pH setpoint is 7.3.
**b)** Agitation via free magnetic stir bar, otherwise via suspended magnetic spinner.
**c)** 1 N NaOH.
**d)** Final inactivated and purified bulk, prior to co-precipitation
**e)** Averages do not include values from lot 3.

**Table V. The Final pH of the Aluminum Hydroxide Co-precipitated Hepatitis A Bulk**

| Lot | Final On-line pH (measured at 8°C) | Off-line pH[a] (measured at 20°C) |
|---|---|---|
| 1 | 5.99 (no buffer) | 5.7 (no buffer)<br>6.6 (with buffer) |
| 2 | 6.60 | 6.3 |
| 3 | 6.35 | 6.2 |
| 4 | 6.55 | 6.3 |
| 5 | 6.42 | 6.1 |
| 6 | 6.77 | -NM- |
| 7 | 6.53 | 6.4 |
| 8 | 6.43 | 6.1 |
| 9 | 6.45 | 6.2 |
| 10 | 6.45 | 6.0 |
| **Avg.[b]** | **6.5 ± 0.1** | **6.2 ± 0.1** |

**a)** Measured with a Fisher probe and Orion meter.
**b)** Average does not include data in lot 1 and 6.

**Table VI. Aluminum Levels in Aluminum Products from Various Sources.**

| Precipitated Aluminum Source | Al (mg/ml) QC labs | Al (mg/ml) Other |
|---|---|---|
| *Demonstration Lots* | | |
| Lot 1 | | 0.476 |
| Lot 2 | 0.45 | |
| Lot 3 | 0.41 | 0.512 |
| Lot 4 | 0.54 | |
| Lot 5 | 0.40 | 0.470 |
| Lot 6 | 0.45 | |
| Lot 7 | 0.43 | 0.446 |
| Lot 8 | 0.40 | 0.443 |
| Lot 9 | 0.47 | 0.431 |
| Lot 10 | 0.48 | |
| **Average =** | **0.45 ± 0.05** | **0.46 ± 0.03** |
| *Test Runs* | | |
| Run #1 | | 0.446 |
| Run #2 | | 0.431 |
| Run #3 | | 0.452 |
| Run #4 | | 0.464 |
| **Average =** | | **0.45 ± 0.01** |
| *Aluminum Hydroxide Diluent* | | |
| Run #1 | | 0.491 |
| Run #2 | | 0.497 |
| Run #3 | | 0.449 |
| **Average =** | | **0.48 ± 0.03** |

**Table VII. The PZC and particle size of aluminum hyroxide adjuvant from various sources.**

| Precipitated Aluminum Source | Point of Zero Charge[a] (in NaCl) | Mean Particle Size[b] (microns) |
|---|---|---|
| *Demonstration Lots* | | |
| Lot 1 | 7.15 | -- |
| Lot 3 | 7.16 | 3.1 |
| Lot 4 | -- | 1.8 |
| Lot 5 | 7.58 | 1.6 |
| Lot 7 | 7.49 | 4.0 |
| Lot 8 | 7.29 | -- |
| Lot 9 | 7.36 | -- |
| **Average =** | **7.3 ± 0.2** | **2.6 ± 1.1** |
| *Test Runs* | | |
| Run #1 | 7.20 | 2.6 |
| Run #2 | 7.37 | 3.7 |
| Run #3 | 7.31 | 4.3 |
| Run #4 | 7.41 | 2.6 |
| **Average =** | **7.3 ± 0.1** | **3.3 ± 0.8** |
| *Aluminum Hydroxide Diluent* | | |
| Run #1 | 7.48 | 4.1 |
| Run #2 | 7.36 | 4.5 |
| Run #3 | 7.03 | 3.0 |
| **Average =** | **7.3 ± 0.2** | **3.9 ± 0.8** |

a) PZC is measured by electrophoretic motility across a pH range.
b) Size is measured on a Microtrac II Particle Size Analyzer with water as the dispersing solution. Solutions are mixed >20 minutes to reach equilibrium particle-size distribution.

**Table VIII. Formaldehyde Clearance in the Inactivated Bulk after Six Decants for 5 Lots**

| Lot | Formalin in Inact. Bulk | Residual Formaldehyde remaining in bulk/supernatants | | | | | | |
|---|---|---|---|---|---|---|---|---|
| | | Decant 1 | Decant 2 | Decant 3 | Decant 4 | Decant 5 | Decant 6 | Final bulk |
| | mcg/ml | | | | | | | |
| 2 | 380 | 100 % | 47 % | 23 % | 9.4 % | 4.2 % | 1.8 % | 0.7 % |
| 3 | 420 | 100% | 45 % | 17 % | 8.6 % | 3.5 % | 1.6 % | 0.6 % |
| 4 | 380 | 100% | 51 % | 23 % | 9.6 % | 6.0 % | 2.8 % | 1.3 % |
| 5 | 390 | 100% | 48 % | 23 % | 11 % | 5.4 % | 2.4 % | 1.2 % |
| 6 | 360 | 100 % | 49 % | 21 % | 10 % | 3.9 % | 2.1 % | 1.1 % |
| Avg. | 390 | 100 % | 48 % | 21 % | 9.8 % | 4.6 % | 2.1 % | 1.0 % |
| Target[a] | | 100 % | 46 % | 22 % | 10 % | 4.6 % | 2.2 % | 1.0 % |

a) The target is the theoretical removal based on a decant volume of 54% of the total volume.

(no antigen), and VAQTA, and are outlined in Table IX. The runs revealed that air entrainment into the recirculation loops can occur at high total recirculation rates. The final operational protocol that was developed included a main loop recirculation rate of 4.0 L/min. (after initial re-suspension), a manifold recirculation rate of 1.5 L/min. and filling at 200 vials/minute; near the end of the fill (<20L), the main loop recirculation was stopped to prevent air entrainment. Simulations were carried out to confirm that the product remained homogeneous at the reduced total recirculation rate. These combined measures ensured that homogeneous product was delivered to the vials throughout the entire reduction in bulk volume, with no air entrainment. Further simulation fills using aluminum hydroxide diluent and RECOMBIVAX HB were carried out at different scales (30-90 L) to further confirm process robustness.

Based on statistical analysis of process and assay data collected during development, specifications were established for final antigen concentration after dilution and filling. Other specifications, including aluminum concentration, pH , and volume-of-fill were set primarily based on predetermined ranges for RECOMBIVAX HB and other aluminum-based products.

**Demonstration and Formal Validation Phase.** As shown in Table II, the critical quality attributes for the dilution process include the concentration of the active ingredient, homogeneity of the product, and sterility. The critical process parameters that affect these attributes include system accuracy (scale, pump, etc.), volume ratio (as discussed earlier), and mixing rate and time. To formally validate this step, samples should be taken throughout the step and assayed for these attributes, according to a pre-approved protocol with pre-determined specification ranges. Initial samples of the concentrated bulk and diluent should be taken prior to the dilution to confirm their homogeneity. After dilution, at least two point samples should be taken from the diluted bulk vessel during mixing and assayed for the critical quality attributes. Preferably, the first of these point samples is taken after the solution becomes homogeneous (based on developmental data), but prior to the end of the mixing operation. A slower mixing speed during testing can also be used to establish a safety margin. These point samples should be taken and assayed for three consecutive lots which span the range of batch volumes expected during manufacturing. Alternatively, a "worst-case" volume approach could be used if the rationale is sound.

Generally, a sterile challenge is required to validate the sterility quality attribute. In this procedure, a growth medium is substituted for the concentrated and diluent bulks, and a mock dilution is carried out following the standard procedures. After the mock dilution and fill, the vials are incubated and checked for growth; the bulk dilution may also be tested. Filling into vials provides an indication of the contamination severity if a failure (i.e. growth) occurs. This worst-case approach is superior to assaying a small number of samples for each batch, and can be applied across different products that use the same process step with the same equipment.

For VAQTA, the post-dilution mixing specification was 30 minutes at 5.6 L/min. Formal validation of this step was performed on the demonstration lots according to a pre-approved validation protocol. Samples were taken from concentrated product and diluent bulks prior to dilution, and from the mixing diluted bulk after 20 and 30 minutes of mixing. Samples were assayed for antigen and/or aluminum, and shown to fall within the pre-determined acceptable ranges. A range of

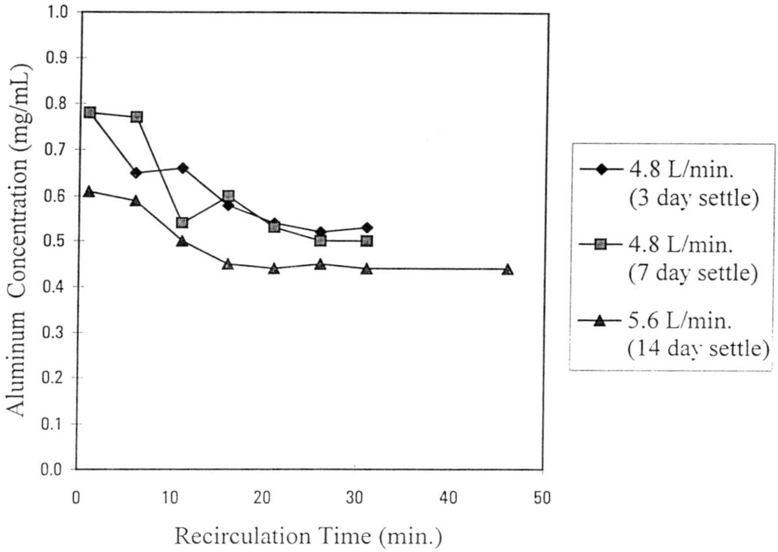


Figure 5. Effect of Recirculation Flowrate on Re-suspension of RECOMBIVAX HB.

**Table IX. Process Development Fills**

| Solution | Main Pump Recirc. Rate (L/min) | Manifold Recirc. Rate (L/min) | Air Entrainment |
|---|---|---|---|
| Water | 5.6 | 1.5 | Yes |
| Alum Diluent | 4.0 | 1.5 | No |
| Alum Diluent | 5.6 | 1.5 | No |
| Alum Diluent | 7.0 | 1.5 | No |
| Alum Diluent | 7.0 | 5.6 | Yes |
| VAQTA | 4.0 | 1.5 | No |

batch sizes was used across three consecutive lots. The sterility of this step for VAQTA was established through a sterile challenge as outlined previously.

The critical quality attributes for filling are the same as those for dilution, with the additional requirement of fill volume. Critical process parameters include batch volume, bulk age time (which can affect compactness for a suspension), mixing throughout the entire system, fill pump settings, and filling speed. Filling validation is normally performed by taking samples of vials or syringe at time-points during the process, and then assaying them for the critical attributes. At least three sample intervals (i.e. beginning, middle, and end of the fill) should be used. In addition to the requirement that all data fall within the pre-determined specifications, no trends in the data (for example, concentrations decreasing through the fill) should be observed. For multiple filling pumps, samples should be taken such that the various pumps and filling needles are represented. A higher filling rate than that planned for normal manufacturing can be used to demonstrate a safety margin, however one must be careful in attempting to use different mixing speeds since there can be a trade-off between product homogeneity and air entrainment. Again, an approach using three consecutive lots which span the range of batch sizes expected during manufacturing, or a worst-case approach, should be used.

For VAQTA, sampling was performed at a minimum of 10 time-points throughout the demonstration fills. Samples were assayed for antigen, aluminum concentration, and checked for fill weight. These results are shown in Figures 6, 7, and 8 for the three consistency lots (lots 6-8) for the purified and formulated bulks. The dotted lines show the pre-determined specification limits; all the validation data falls well within these specifications. This study also linked the filling validation with the dilution validation study by using the same three consecutive batches for both studies. A worst case approach with respect to batch volume was not used; the batch volumes spanned the normal volumes expected (30-60 L). Sufficient data from simulation runs with aluminum hydroxide diluent and RECOMBIVAX HB exists, which confirms that batch volumes as high as 90L could be accommodated. Finally, the sterility of this step for VAQTA was established through a satisfactory media challenge.

## Current and Future Trends in Sterile Processing -The Validation Perspective

The rapid advancement in technology over the last two decades has made validation particularly difficult to manage. Reliance on computers and automation (i.e. electronic batch records) has increased the validation burden. To stay current, companies have dedicated considerable time and resources to validation; its cost may have been as high as $390 million in the pharmaceutical industry in 1995 alone (*15*). To prevent costs from spiraling out of control, companies should use good judgment when designing a validation program and focus on the critical aspects of the process. Companies will have to rely more heavily on vendors to provide new technologies that are validatable and straightforward to scale up. Finally, the industry and the FDA need to establish clearer and simpler standards to allow companies to comply with regulatory requirements, thoroughly and economically. The perspective that validation is "documented common sense" is a good one to follow (*2*), but will not suffice in the increasing complex bio-pharmaceutical industry.

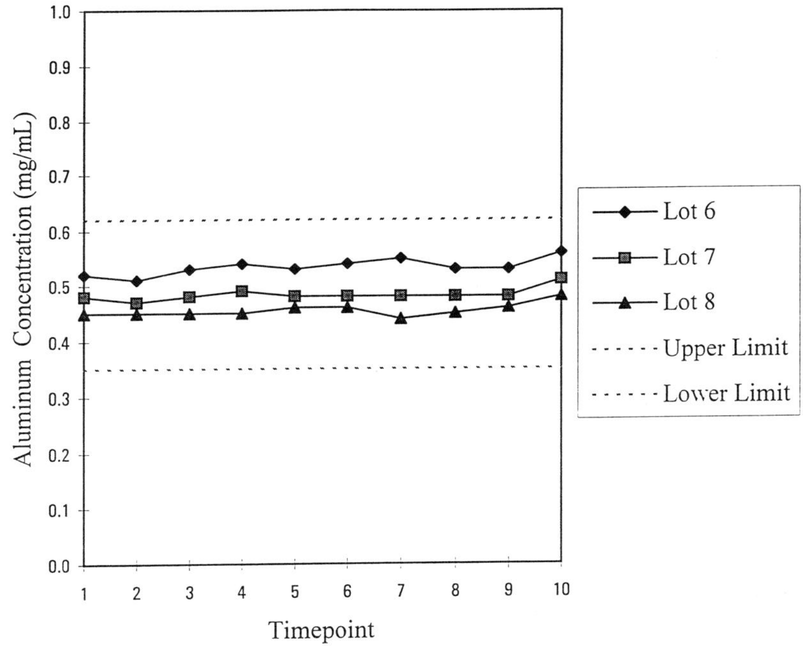


Figure 6. Validation Results for VAQTA: Aluminum Levels across the Fill.

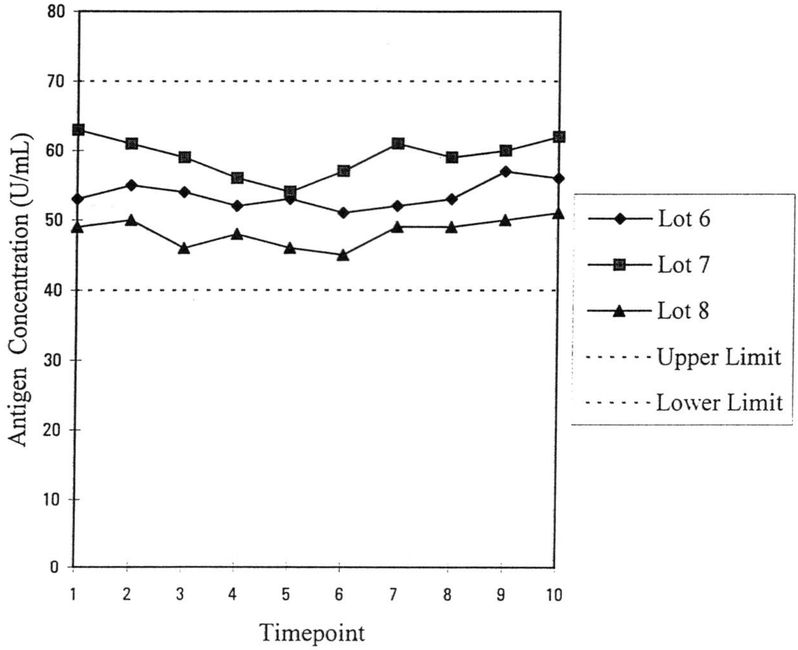


Figure 7. Validation Results for VAQTA: Antigen Levels across the Fill.

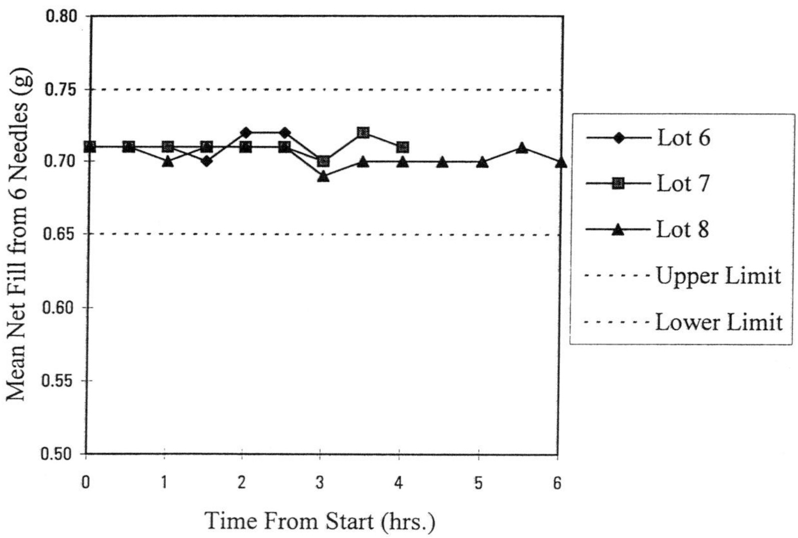


Figure 8. Validation Results for VAQTA: Vial Fill Weights across the Fill.

In the arena of sterile processing, several technological advancements have been made to reduce contamination risks, including locally controlled environments for filling and other sterile processing, new vessel closures for small-scale glass vessels, sterile welder technology, and new CIP/SIP systems for closed systems. Mixing is also key to consider when designing a new sterile process. Assuring that these new processes and systems function as designed requires a considerable validation effort up-front. This can be a major short-term cost for an often worthy long-term investment.

Locally controlled environments (LCE) or barrier systems represent a significant advancement in sterile filling technology by eliminating people from the filling environment. The interior of these systems is pre-sterilized using vaporized hydrogen peroxide or some other agent prior to a filling run, and access to equipment during a run is through glove-ports. One challenge for these systems includes the transfer of components in and out of the barrier after the main filling environment has been sterilized. LCEs have the potential to greatly improve the sterility assurance level (SAL) achievable during the filling operation.

Emphasis on container closure integrity has increased over recent years. This has arisen from the need to validate the sterility of closed vessels. Several methods have been employed to accomplish this: helium leak, pressure-hold, media challenge, and microbial submersion. The latter has been imposed recently as a more direct challenge to the closure; many companies, including Merck, have accepted it in their validation programs. On the whole, however, validation testing has identified several suitable closures for use with glass vessels (100 mL-50 L). A future advancement for these closures is to couple them with tubing that can be sterile-welded to maintain a closed system during transfers and sampling.

As described earlier, sterile welding provides an effective way to make sterile connections between silicone-type tubing lines. The welder effectively allows connections to occur in a closed system, making it is possible to maintain sterility even when connections are made in an unclassified environment. This technology is especially useful when processing small batches, which is often the case in research and development, but will have wide-spread use throughout the industry with the emergence of larger-sized tubing in the near future. Sterile welding provides considerable flexibility and speed for sampling and product transfers compared to the current SIP systems that are available today.

Advances in clean-in-place/steam-in-place technology have allowed equipment and piping to be cleaned and steamed without moving equipment to cleaning areas or autoclaves. Facilities and equipment are designed and constructed to contain connections to WFI and clean steam, and to allow total coverage of the internal equipment area. Already under widespread use throughout the industry today, these new designs allow manufacturing equipment to be sanitized or sterilized in a closed system without exposure to the outside environment.

Finally, another key consideration for sterile operations is mixing. This is particularly important for sterile suspensions, such as aluminum-based adjuvants. Although this paper illustrates how recirculation mixing is effective for creating and maintaining a homogeneous suspension, impeller mixing is also commonly employed in sterile operations. The choice of mixing will depend on the situation; trade-offs exist between the different approaches. Impellers (with baffles) are recommended for applications where mixing is particularly important, such as reactions (e.g. co-precipitation) and re-suspension of highly aggregated solids. The main disadvantages of impellers, however, are leaks at the seals, particle generation, the inability to mix adequately at low volumes, and cleanability of the impeller and baffles, especially underneath if a CIP system is used. Alternatively, magnetic-driven impellers can also be employed to avoid some of these issues. Baffles are not commonly utilized in the biotech industry (unless critical to good mixing) due to cleaning issues. Some recirculation mixing, on the other hand, is always required during filling of suspensions, regardless of the mixing approach chosen for the tank. The suspension must be recirculated through the filling manifold to prevent product from settling out. As the product level drops during filling, recirculation is also effective for maintaining product homogeneity until the tank is emptied.

**Acknowledgments**

The authors would like to thank Mike Turrisi for his engineering expertise in designing and procuring the co-precipitation vessel and pH automation system, Susan Isabell and others in the Bioanalytical Group for their considerable analytical support, the production staff for VAQTA, Prof. Stanley Hem and his laboratory at Purdue University for adjuvant testing and consultation, and the BioQC laboratories at Merck for other analyses. Their efforts contributed significantly to the success of this project. Special thanks is also extended to Dr. Ann Lee for her thorough review of the manuscript.

**References**

1. Food and Drug Administration, Code of Federal Regulations, Part 211.113 (b), Validation of sterilization processes, Office of Federal National Archives and Records Administration, Washington DC, 1989.
2. Chapman, K. G., "A History of Validation in the United States- Part I", *Pharm. Technol.*, **1991**, *15, no. 10 (October),* pp.82-96
3. Lee, J.L., In *Pharmaceutical Process Validation;* Berry, I.R.; Nash, R. A., Eds.; Marcel Dekker: NY, NY, 1993, 2nd Ed., Chap. 19; pp. 573-586.
4. Akers, MJ.; Anderson, N.R., In *Pharmaceutical Process Validation;* Berry, I.R.; Nash, R. A., Eds.; Marcel Dekker: NY, NY, 1993, 2nd Ed., Chap. 3; pp.25-87.
5. Hofmann, K.L. Jr., In *Pharmaceutical Process Validation;* Berry, I.R.; Nash, R. A., Eds.; Marcel Dekker: NY, NY, 1993, 2nd Ed., Chap 4; pp. 89-166.
6. Avis, K.E., In *Remington's Pharmaceutical Sciences*; Remington, J. P.; Gennaro, A.R Eds.; Mack Publishing Co.: Easton, Pa., 18$^{th}$ Ed, 1990; Chapter 84, pp. 1545-1569.
7. VAQTA and RECOMBIVAX HB are registered trademarks of Merck & Co..
8. Hagen, A.J.; Aboud, R.A.; DePhillips, P.A.; C. N. Oliver; Orella, C.J.; Sitrin, R.D. *Biotechnol. Appl. Biochem.*, **1996**, *23*, pp.209-215.
9. Sitrin, R.D.; Wampler, D.E.; Ellis, R.W. In *Hepatitis B Vaccines in Clinical Practice*, Ellis, R. W., Ed.; Marcel Dekker: NY, NY, 1993, pp. 83-101.
10. Glenny, A.T.; Pope, C.G.; Waddington, H.; Wallace, U. *J. Pathol. Bactieriol*, **1926**, *29*, p. 38.
11. Hem, S.L.; White, J.L. *J. Parenteral Sci. Tech.*, **1984**, *38, no.1,*p. 200.

12. Hem, S.L.; White, J.L. In *Vaccine Design*; Powell, M.F.; Newman, M.J., Eds., Pharmaceutical Biotechnology; Plenum Press: NY, NY, 1995, Vol. 6; pp. 249-76.
13. Mims C.A., Playfair J.H.L., Roitt I.M., Wakelin D., Williams, R. *Medical Microbiology*; Mosby: St. Louis, 1993; Chap. 36, p. 36.1-36.20.
14. Shirodkar, S.; Huthcinson, R.L.; Perry, D.L.; White, J.L.; Hem, S.L. *J. Pharm. Res.*, **1990**, *7, no.12*, pp. 1282-8.
15. Langer, L.J., *Genetic Engineering News*, **1997**, *Vol. 17, no.11(June 1)*, p.1.

Chapter 12

# Changes in Biologics Regulations: Impact on the Development and Validation of the Manufacturing Processes for Well-Characterized Products

**Kimberlee K. Wallace[1] and Antonio R. Moreira[2]**

**Departments of [1]Continuing Education and [2]Chemical and Biochemical Engineering, University of Maryland, Baltimore County, Baltimore MD 21250**

The changes in the FDA regulations governing biologics will provide biopharmaceutical manufacturers enhanced flexibility in the development and implementation of manufacturing processes. The use of pilot facilities for production of Phase III clinical products will be permitted provided the clinical product is "comparable to the marketed product. Proving such "comparability" will necessitate increased biomolecular characterization potentially resulting in the development and validation of new analytical assays. Additionally, demonstrating process and product consistency between the pilot and production scales will place a greater emphasis on process validation, especially at the bulk stages. Although the new regulations are positively received by the industry, the impact and potential costs associated with their implementation are not yet well defined. This paper investigates the impact of these regulations on bioprocess development and validation for well-characterized biologics.

Regulation of biologics manufacturing commenced following the production of the diptheria antitoxin using an equine system. Unfortunately, the horse used was contaminated with tetanus resulting in the untimely death of several people to whom the antitoxin was administered *(1)*. Based upon this episode, the Food and Drug Administration (FDA) developed a stricter set of regulations for biological products than those in place for drug products in an attempt to ensure product quality and protect the health of the public. Control of the manufacturing process from the onset was expected by the Agency due to the perceived lack of ability to characterize the final product in terms of purity, identity and freedom from adventitious agents. Although biotechnology-derived products have since demonstrated an excellent safety record, the FDA has been slow to modify its existing regulations, resulting in an expensive and time-consuming path to market for new biopharmaceutical products.

The approval process for biological products differed from that of traditional drug products. As for drug products, companies were required to submit an investigational new drug application (IND) prior to studying the product in human subjects. Successful completion of the clinical trials required that companies simultaneously submit a product license application (PLA) and an establishment license application (ELA), both of which required approval prior to commercial marketing and distribution.

With the Reinventing Government Initiative of 1995 *(2)*, the FDA streamlined the regulations making the approval process for certain biotechnology-derived products parallel that of traditional pharmaceuticals. For well-characterized products including rDNA-derived therapeutics, monoclonal antibodies, synthetic peptides containing fewer than 40 amino acids and therapeutic DNA plasmid products *(3)*, the ELA and PLA have been collapsed into a single document, the Biologics License Application (BLA).

Complying with the FDA regulations requires companies make a significant financial investment in the development of potential drug candidates before the success of the product is guaranteed. Continued financial investments are required throughout the lifetime of the molecule to maintain compliance and to reproducibly manufacture a product meeting predetermined quality attributes. The FDA is cognizant of the concerns of the industry and has introduced several new regulations pertaining to the use of pilot facilities (*4*), demonstrating product comparability (*5*), making changes to an approved application (*6,7*) and defining well-characterized products (*8*). Theoretically, these regulations will benefit the industry by reducing the time required for product approval. However, the economic impact of these regulations on the manufacturers has not been examined. The objectives of this work were several-fold: to determine the costs associated with development of the manufacturing process and associated assays used in product characterization and to examine the economic impact of the recent changes in the FDA regulations on biopharmaceutical manufacturers.

## Methodology

A survey was sent to a representative sample (25) of small/medium and large companies in the United States involved in the manufacture or development of biopharmaceutical products requesting information on their process development and associated validation costs. Follow-up phone interviews were conducted with companies which responded to the survey as well as a random selection of companies from which no response was received.

The survey was divided into three sections. The first section requested information on the size of the company, the number of employees involved in process development, validation and regulatory activities and the number of biopharmaceutical products currently approved as well as those in clinical trials. In analyzing the data from the survey, information was often grouped based upon the size of the company. Responding companies (3 respondents) with fewer than 3500 employees were classified as small/medium companies while companies (4 respondents) employing more than 3500 people were considered large companies. The second section concentrated on development and validation of assays used for characterization of biotechnology-derived products. Information on process development and validation activities including the perceived impact of the new regulations on their organization was collected in section three. The costs calculated in this study include the costs of personnel, supplies and equipment depreciation.

## Results and Discussion

**Process Development.** In the manufacture of any pharmaceutical product, gaining market share is critical to the future success of the product, and in many cases, the company. However, this rush to market must be balanced by an adequate period of development and testing to assure that the product is both safe and efficacious. As shown in Table I, the amount of time spent in process development varies with the size of the company.

**Table I: Years Spent in Development Before Entering the Clinic and Associated Costs**

| *Company Size* | *Years* | *Cost* |
|---|---|---|
| Small/medium | 1-2 | $230-460,000 |
| Large | 2-4 | $460-860,000 |

Surprisingly, our results indicated that the larger companies spent more time on development activities before proceeding to the clinic than their smaller counterparts. One reason for this difference may be that larger companies wish to refine their process to a high degree before entering the clinic. Although this increased development time increases the costs associated with bringing a product to market, it may minimize the number of surprises encountered during clinical trials in light of the stronger knowledge about the production process, thereby, resulting in a significant cost savings in the long run. The shorter development times used by the smaller companies may be due to economic constraints. To them, reducing the development time by several years may result in a cost savings of $400,000. Since less than 10% of the products which enter clinical trials are successful, they may wait to refine their process until after the completion of Phase I or II clinical trials, when the success of the product is more assured allowing them to more likely recoup their losses. The reduced development time also permits them to enter the clinic sooner than their competitors. If their product successfully completes clinical trials, they will gain market exclusivity, a lucrative position. It was interesting to note that the development times for the large companies ranged over a 2-4 year period. Our results indicated that the longer development times were associated with companies which did not currently have a licensed biologic on the market, possibly due to the lack of full familiarity with the expectations of the Center for Biologics Evaluation and Research (CBER).

**Assay Development and Validation.** In recent years, advances in instrumentation have resulted in enhanced assay capabilities providing for more accurate characterization of product quality. From the perspective of the FDA, the more complete the characterization of both process intermediates and final product, the less likely that a product which threatens the safety of the public will reach the marketplace. Survey of the industry revealed that small to medium size companies typically conduct 15-20 assays per product to characterize in-process intermediates and final product. In contrast, larger companies typically perform 15-40 assays per product. In both cases, the total number of assays conducted depended on the system used for expression of the product of interest. The use of mammalian expression systems required a larger number of assays due to concern over the presence of adventitious agents.

Although assays may be based on diverse product characteristics including absorption properties, protein chemistry or metabolic studies, our survey indicated that the assays were largely chemical/biochemical, immunological and biological in nature. As seen in Table II, the development and validation times vary significantly depending upon the type of assay. Biochemical/chemical assays which require 450-720 person-hours to develop and validate account for 70% of the assays performed for product characterization. Bioassays, on the other hand, require twice as many person-hours for development and validation and account for approximately 15% of the total number of assays performed.

**Table II: Person-Hours and Costs Required For Development and Validation of Assays Used in the Characterization of Biopharmaceutical Products**

| *Assay Type* | *Person-Hours* | *Cost/assay* |
|---|---|---|
| Immunological | 360-500 | $43-60,000 |
| Chemical/Biochemical | 450-720 | $54-86,000 |
| Bioassay | 900-1200 | $108-144,000 |

**Case Study.** To fully appreciate the costs associated with bringing a product to market, it is informative to put the costs in perspective. Using the costs calculated above and assuming that a biopharmaceutical product requires 20 assays for characterization and release, Table III shows total costs associated with assay development and validation.

**Table III: Assay Development/Validation Times and Costs**

| *Assay Type* | *Total Number of Assays* | *Total Cost* |
|---|---|---|
| Immunological | 3 | $129,000-180,000 |
| Chemical/Biochemical | 14 | $746,000-1,200,000 |
| Bioassay | 3 | $324,000-432,000 |
| Total | 20 | $1,200,000-1,800,000 |

Table IV summarizes the typical costs associated with the development of a biopharmaceutical product. In addition to investing upwards of a million dollars in developing a reproducible manufacturing process, manufacturers are required to characterize their cell banks in order to ensure genetic stability and freedom from adventitious agents (*9*). Characterization costs range from $10,000 for a microbial system to $30,000 for a mammalian expression system. These figures represent only the costs of performing previously validated assays, either in-house or by a contract laboratory. Inclusion of validation costs would significantly increase this figure.

The cost estimate of approximately 1.7 to 2.7 million dollars does not include associated process validation costs since the industry was not willing to share these numbers. Developing a new product is a gamble. Bringing a product to clinical trials can cost the company over three million dollars with no guarantee that their product will be successful, allowing them to recoup their costs.

**Table IV: Cost Summary**

| | | ***Total Cost*** |
|---|---|---|
| Process Development | | $460,000-860,000 |
| Cell Line Characterization | | $10,000-30,000 |
| Assay Development/Validation | | $1,200,000-1,800,000 |
| | Total Cost | ≈$1,700,000-2,700,000 |

**Use of Pilot Facilities.** Prior to 1995, companies were required to manufacture phase II clinical materials at the final commercial scale. This ruling required companies to build their commercial scale manufacturing facility without an assurance that their product would reach the market. If the product was unsuccessful during the phase III clinical trials, the organization was left with a huge financial loss, or in the case of smaller, companies, possibly bankruptcy. In response to these concerns, the FDA changed its ruling concerning the use of pilot facilities. Rather than requiring that phase III materials be manufactured in a commercial facility, the Agency permitted the licensing of pilot facilities (*4*). This change meant that companies could produce their clinical material at a smaller scale and ultimately license this smaller facility for commercial manufacturing. By not requiring the construction of a new larger-scale facility, this change could save a company from $25-50,000,000. However, as shown in Table V, not all companies plan to take advantage of this rule change.

**Table V: Percentage of Companies Planning to License a Pilot Facility for Commercial Manufacturing**

| *Company Size* | *Yes* | *No* |
|---|---|---|
| Small/medium | 100% | 0% |
| Large | 0% | 100% |

In general, as seen in Table V, this flexibility in commercial manufacturing will be advantageous to the small to medium size companies which do not have the capital available to build a facility which may sit idle if negative results are obtained in the clinic. In contrast, the larger companies which typically have greater access to
capital do not seem to plan to exercise this flexibility. Rather, they will likely continue to build or modify a commercial facility prior to the completion of phase III clinical trials.

**Changes to an Approved Application.** Under the previous system of regulations, companies were not permitted to make modifications to the manufacturing processes of approved products without prior approval from the FDA. Due to the large number of applications awaiting review and the inadequate number of FDA personnel, companies would wait months for changes to be approved. Benefiting from current manufacturing experience, the FDA modified the regulations to permit certain types of process changes to be implemented without prior Agency approval (*6,7*). In general, the industry views this change favorably. Although this change will not affect the validation costs associated with process modifications, it will encourage companies to continually improve the manufacturing processes for those products which are already on the market. Quantification of these improvements is difficult, but it can be anticipated that they will result in lower production costs and increased production capabilities. In the long run, such savings could be passed on to the consumer in the form of lower drug prices.

**Demonstration of Product Comparability.** Previously, the FDA often required that approved products produced following changes in the manufacturing process undergo a new series of clinical trials to prove their safety and efficacy. However, improvements in product characterization methods have helped the FDA to modify their position. In a recent guidance document (*5*), the Agency states:

> "The ability of the manufacturer to use validated and sensitive assays to demonstrate a product's identity and structure, biological activity and clinical pharmacology will provide a basis for determining whether product comparability can be established without repeating efficacy studies."

This statement suggested to us that industry might increase the number of assays conducted on a product to facilitate demonstration of comparability. However, all the companies surveyed insisted that this change would not impact the number and types of assays used for characterization. They felt that the number of assays which they performed was adequate for ensuring product quality and that this number would not need to increase to adequately document product comparability. Although this change in thinking of the Agency does not modify the costs associated with product characterization or assay validation, the industry enthusiastically embraced this change since it potentially saves them the costs associated with a phase III clinical trial and may allow a comparable product to reach the marketplace more quickly.

**Elimination of Lot Release.** In accord with its strict regulatory control of biologics, the FDA would not typically permit the distribution of biopharmaceutical products prior to release by the Agency. This regulation was a burden to both the industry and the Agency. Depending upon the nature of the product, the market value of samples shipped to the FDA for testing could be in the range of $20-50,000 per lot

In terms of economics, elimination of the requirement for lot release for well-characterized products will benefit the industry in two ways. Firstly, samples that used to be sent to the Agency will now be available for commercial sale resulting in increased corporate revenue. Secondly, the time required for product release will be shorter. For new products, this means that they will reach the marketplace sooner and possibly capture an increased share of the market.

**Elimination of the Establishment License Application.** The most sweeping change in the regulation of biologics is the elimination of the establishment license for the facility manufacturing biopharmaceutical products. Unlike traditional drug products, biologics were originally granted a license for the product and a separate license for the manufacturing facility. This system fostered an interactive dialogue between the industry and the Agency prior to the pre-approval inspection. In the new regulations, which more closely follow those of drugs, the majority of information on the facility is fathered during the pre-approval inspection. Although industry typically favors this change, it agrees that it may negatively impact certain companies.

As shown in Table VI, some members of the industry feel that the approval of biological products may be delayed because of facility-related issues. In the previous system, these issues could have been addressed in the establishment license application and through an interactive dialogue with the FDA prior to the pre-approval inspection. In general, the large companies were less concerned about these changes than the smaller companies, most likely due to their familiarity with the expectations of the FDA. Smaller companies, which lacked experience with inspections, were concerned about these changes which could result in approval delays and possibly loss of revenue and market share.

**Table VI: Increase in Approval Times for BLA**

| *Company Size* | *Yes* | *No* |
|---|---|---|
| Small/medium | 100% | 0% |
| Large | 33% | 67% |

## Summary

In summary, bringing a new biopharmaceutical product to market is expensive, requiring the investment of two to three million dollars before the product is tested in human volunteers. The changes to the FDA regulations are a positive step in the minds of the manufacturing community. By decreasing the hurdles involved in gaining product approval, the Agency is encouraging processing improvements and innovations. In the coming years, this creativity may result in decreased production costs and ultimately lower priced biopharmaceutical products.

## Literature Cited

1. Gamerman, G.E., Mackler, B.F., and Landa, M.M. *Food, Drug, Cosmetic and Medical Device Law Digest.* **1997.** *14*, pp. 9-14.
2. FDA Policy Report. Reinventing the Regulation of Drugs Made from Biotechnology. November **1995.**
3. *Federal Register.* **1996.** *61,* pp. 24227-24233.
4. Food and Drug Administration. FDA Guidance Document Concerning the Use of Pilot Facilities for the Development and Manufacture of Biological Products. **1995.**

5. Food and Drug Administration. FDA Guidance Concerning Demonstration of Comparability of Human Biological Products, Including Therapeutic Biotechnology-Derived Products. **1996.**
6. Food and Drug Administration. Changes to an Approved Application for a Well-Characterized Therapeutic Recombinant DNA-Derived and Monoclonal Antibody Biotechnology Product; Draft Guidance. **1996.**
7. Food and Drug Administration. Changes to an Approved; Draft Guidance. **1996.**
8. *Federal Register*. **1995.** *60,* p. 63048.
9. Food and Drug Administration. Points to Consider in the Characterization of Cell Lines Used to Produce Biologicals. **1985.**

# Indexes

# Author Index

# Subject Index

R